Essays in Biochemistry

volume 35 2000

Essays in Biochemistry

Molecular Motors

Edited by G. Banting and S.J. Higgins

Portland Press

Essays in Biochemistry is published by Portland Press Ltd
on behalf of the Biochemical Society

**Portland Press
59 Portland Place
London W1N 3AJ, U.K.
Fax: 020 7323 1136; e-mail: editorial@portlandpress.com
www.portlandpress.com**

In North America orders should be sent to Princeton University Press,
41 William Street, Princeton, NJ 08540, U.S.A.

British Library Cataloguing-in-Publication Data
A catalogue record for this book is available from the British Library

ISBN   1 85578 103 4
ISSN   0071 1365

Typeset by Portland Press Ltd
Printed in Great Britain by Information Press Ltd, Eynsham, U.K.

Contents

1   **Introduction: the mechano-biochemistry of molecular motors**
Rob Cross

2   **F_1-ATPase: a highly efficient rotary ATP machine**
Kazuhiko Kinosita, Jr., Ryohei Yasuda and Hiroyuki Noji

3   **Motors in muscle: the function of conventional myosin II**
Clive R. Bagshaw

10 Translational elongation factor G: a GTP-driven motor of the ribosome

Wolfgang Wintermeyer and Marina V. Rodnina

11 How do proteins move along DNA? Lessons from type-I and type-III restriction endonucleases

Mark D. Szczelkun

12 Molecular motors in the heart

Fergus D. Davison, Leon G. D'Cruz and William J. McKenna

13 The roles of unconventional myosins in hearing and deafness
Richard T. Libby and Karen P. Steel

Preface

Biological systems abound with examples of molecular motors, biological machines for converting the chemical energy of ATP into mechanical movement by cells.

Apart from the classic examples of muscle contraction and the movement of cells by means of flagella or cilia, molecular motors are responsible for such diverse cellular functions as organelle movement within cells, chromosome segregation during cell division, the intracellular targeting of proteins, and the transport of secretory vehicles, with their cargoes of proteins or neurotransmitters, to the plasma membrane. Gene expression and replication also rely on molecular movement, for instance polymerases process along DNA and ribosomes are translated codon-by-codon along mRNA.

Until recently, interest in molecular motors has been the province of a select group of biologists. However, all that changed with the award of the Nobel Prize for Chemistry to John Walker and Paul Boyer in 1997 for their work in elucidating the mechanism of the rotary generator of ATP in the mitochondrion, the F_1-ATPase. This brought the exciting developments in the field of molecular motors to the attention of the whole scientific community and, via the media, to a wider world audience.

It is timely to have this volume, covering as it does many of the most exciting developments concerned with molecular motors. All of the main motor systems are featured, and the recent advances obtained by the application of gene-cloning methods are described. There are also readable descriptions of how basic research in biological motor mechanisms is shedding light on important human disorders, such as hearing loss and heart disease.

The volume commences with a succinct survey by Rob Cross, after which Kazuhiko Kinosita's group presents an exciting description of the F_1-ATPase. Three essays on myosins then follow — by Clive Bagshaw, by Georg Kalhammer and Martin Bähler, and by Alex Knight and Justin Molloy. Jon Kull's description of the kinesins leads into an essay by Alistair Harrison and Stephen King on the dynein family of motor proteins. Dawn Signor and Jonathan Scholey then explain microtubule-based transport within cells and David Woolley describes the motors of cilia and flagella. An important translational motor of the ribosome is discussed by Wolfgang Wintermeyer and Marina Rodnina, followed by a description by Mark Szczelkun of how type-I and type-III restriction enzymes move along DNA. Finally, Bill McKenna, Fergus Davison and Leon D'Cruz explain molecular motors in the heart and Richard Libby and Karen Steel address myosins in hearing and deafness.

We thank all our authors for their very fine essays and for addressing the brief we gave them in such exciting ways. We hope the volume will be attractive to senior undergraduates and junior postgraduate students who wish to learn more about molecular motors and to participate in the exciting discoveries that lie waiting to be made in the opening years of the new century and millennium. Our thanks also go to Sophie Dilley and to Portland Press for their work to ensure the high-quality production of this book.

George Banting (Bristol) and
Steve Higgins (Leeds)
December 1999

Authors

Kazuhiko Kinosita, Jr. obtained his BSc in Physics in 1969 and his PhD in 1976 from the University of Tokyo. He constructed a nanosecond time-resolved fluorometer, designing, machining, soldering and writing the software himself. He is thus basically a toy maker. He then spent 2 years at Johns Hopkins University Medical School, where he found that the application of an electric field across a cell results in the formation of small pores in the cell membrane (now called electroporation), that the pore size can be controlled, and that the membrane can be resealed. In 1978, he moved to the Institute of Physical and Chemical Research (RIKEN), and studied dynamic structures of membranes, muscle proteins and DNA through time-resolved fluorescence/phosphorescence depolarization measurements. In 1989, he became Professor of Physics at Keio University. Before that, in 1987, he decided that he would concentrate on optical microscopy. He invented a pulsed-laser microscope to image electroporation at microsecond intervals. He made a multi-view microscope to study cellular events during fertilization. His group at Keio has also studied actomyosin motors extensively using optical tweezers and single-fluorophore imaging, and has produced giant liposomes and demonstrated rapid formation of protrusions in liposomes containing actin. Recent topics include characterization of the F_1 motor and tying knots in molecular strings. The ultimate goal of his group is to help establish a new field of science, 'single-molecule physiology'. **Hiroyuki Noji** received his PhD in 1997 from the Tokyo Institute of Technology. He is currently a postdoctoral fellow of "CREST" Team 13 organized by Dr. Kinosita. His research has concerned the energy transformation between the proton-motive force and phosphoryl potential of ATP by ATP synthase. **Ryohei Yasuda** was born in Japan in 1971. Since receiving his PhD in Physics from Keio University in 1998, he has worked as a postdoctoral fellow with Dr. Kinosita in "CREST". He has studied protein motors, including actomyosin and F_1-ATPase, using optical microscopes.

Clive Bagshaw is Reader in Biochemistry at Leicester University and held a Royal Society Leverhulme Trust Senior Research Fellowship in 1998. He has investigated the kinetic mechanism of the actomyosin ATPase for over 25 years using rapid-reaction methods, fluorescence and electron paramagnetic resonance spectroscopy. More recently he has set up laser-induced fluorescence microscopy for following the myosin ATPase at the single-molecule level. His international collaborations have included study leave periods at Stanford University, California, USA, and Teikyo University, Japan, to develop motility assays *in vitro* for isolated proteins and myofibrils.

Martin Bähler studied Biology at ETH Zürich, Switzerland, where he received his PhD. He was postdoctoral fellow and then Assistant Professor at the Rockefeller University, New York, USA, and group leader at the Friedrich-Miescher Laboratory of the Max-Planck Society, Tübingen, Germany and then at the Adolf-Butenandt-Institute, Ludwig-Maximilians University, München, Germany. He is currently Professor at the Institute of General Zoology and Genetics, Westf.-Wilhems University, Münster, Germany. He has studied contractile proteins of muscle, molecular mechanisms of neurotransmitter release and unconventional myosin. Current research interests include cellular functions of unconventional myosins, molecular mechanisms of force generation, coupling of signal transduction and actin organization. **Georg Kalhammer** studied biochemistry at the University of Tübingen, Germany. For his diploma and PhD work he joined the group of Martin Bähler at the Friedrich-Miescher-Laboratory of the Max-Planck Society in Tübingen. After working on the intracellular localization of the unconventional class-I myosin Myr 4 he switched to the class-IX myosin Myr 5. The studies focused on the function of the class-IX-specific myosin head insertion at the F-actin-binding region of Myr 5. Currently he is working for a law firm specialized in patent protection of biotechnological and pharmaceutical inventions.

Alex Knight is a postdoctoral researcher in Justin Molloy's laboratory at the University of York, where he uses single-molecule fluorescence and optical-tweezers techniques to study the mechanism of force production by myosin. In his previous position at the Whitehead Institute, Cambridge, Massachusetts, USA, he held an EMBO long-term fellowship. His doctoral studies were at the Medical Research Council's Laboratory of Molecular Biology in Cambridge, UK. **Justin Molloy** is a Royal Society University Research Fellow in the Department of Biology, University of York. He originally trained as a biologist and for his DPhil studied the contraction kinetics of single insect flight-muscle fibres. Part of this involved building a mechanical apparatus to measure the micronewton forces produced by *Drosophila* muscle fibres. In 1991 he returned to the UK after a NATO post-doctoral fellowship in Vermont, USA, to develop the optical-tweezers apparatus used to measure the piconewton forces produced by single molecules.

F. Jon Kull is a research-group leader in the Department of Biophysics at the Max-Planck Institute for Medical Research in Heidelberg, Germany. He received his PhD in 1996 from the University of California, San Francisco, USA, where he worked on the crystal structure of the human kinesin motor protein in the laboratories of Robert Fletterick and Ron Vale. His current research group employs X-ray crystallographic techniques to study the structure–function relationships in kinesin and myosin motor proteins, as well as other force-producing nucleotide hydrolases.

Stephen King earned a BSc with honours in Microbiology from the University of Kent at Canterbury, UK, in 1979. He then moved to Jeremy Hyams' laboratory at University College London where he studied the mechanics of mitosis in yeast and was awarded his PhD in 1982. As a Senior Research Associate in the laboratory of George Witman at the Worcester Foundation for Experimental Biology in Shrewsbury, Massachusetts, he studied the Biochemistry and Molecular Biology of *Chlamydomonas* flagellar dyneins. In 1993 he spent several months at Palmer Station, Antarctica, as part of Bill Detrich's group studying dyneins from fish that live at sub-zero temperatures. Since 1993, Dr. King has continued his molecular analysis of dynein components in the Biochemistry Department at the University of Connecticut Health Center in Farmington where he is currently an Associate Professor of Biochemistry and Associate Director of the Genetics, Molecular Biology and Biochemistry graduate programme. **Alistair Harrison** was awarded a BSc with honours in Biology by the University of East Anglia, UK, in 1987 and as an undergraduate working with Richard Warn helped elucidate the timing of microtubule post-translational modification in *Drosophila* embryogenesis. In 1993 he received his PhD from University College London where he worked in Jeremy Hyams' laboratory investigating tubulin post-translational modifications in the nutritive tubes of two hemipteran insects. On moving to the US, he joined Stephen King's laboratory at the University of Connecticut Health Center as a postdoctoral research fellow examining dynein light chains from *Chlamydomonas*. Recently, he moved to Ohio State University to study protein–RNA interactions with Lee Johnson.

Jonathan M. Scholey is a Professor of Cell Biology in the Section of Molecular and Cellular Biology at the University of California at Davis. He obtained a BSc (first class honours) in Cell Biology at the MRC Cell Biophysics Units, Kings College, London, in 1977, and a PhD in Molecular Biology at the MRC Laboratory of Molecular Biology, University of Cambridge, in 1981. He then undertook postdoctoral research on microtubules and the mechanisms of mitosis in the laboratory of Dr J.R. McIntosh at the University of Colorado. He currently works on the functions and mechanisms of action of microtubule-motor-based transport systems in mitotic spindles of *Drosophila* and sea urchin embryos, and in ciliated chemosensory neurons of *Caenorhabditis elegans*. He is the former chair of the Gordon Research Conference on motile and contractile systems. He teaches undergraduate classes in Cell Biology and Biochemistry and a graduate class in Cellular Biochemistry. **Dawn Signor** is a graduate student in the section of Molecular and Cellular Biology at the University of California at Davis, working in the laboratory of Dr. Jonathan Scholey. Dawn obtained a BSc degree in Biology at California State University, Hayward, in 1993, and an MSc degree in Molecular Biology at the same institution in 1996 where she worked on the role of extracellular-matrix molecules in early sea urchin development. In her graduate work at UC Davis, Dawn has studied the role of microtubule-based transport required for the assembly, maintenance and function of non-motile sensory cilia in the *C. elegans* chemosensory nervous system.

David Wooley graduated in Veterinary Science from the University of Liverpool. His research into the physiology of spermatozoa began under the guidance of Alan Beatty at the Institute of Animal Genetics, University of Edinburgh, where he was awarded his PhD in 1969. Postdoctoral appointments at the University of Washington and at Harvard Medical School allowed him to learn electron microscopy under expert tutelage. Subsequently he has held lectureships in the Veterinary Schools at the Universities of Edinburgh and Bristol, where he has contributed to the teaching of the veterinary pre-clinical sciences. His research work has always been focused on the problem of understanding how movement is generated in the eukaryotic flagellum, specifically the sperm flagellum. One consistent theme in the research has been his attempt to use rapid-arrest cryo-electron microscopy as a possible means of reconciling the study of the ultrastructure (e.g. of axonemal dyneins) with the study of living dynamics.

Wolfgang Wintermeyer studied Chemistry at the University of Munich, Germany, where he received his PhD in 1972 for work on tRNA. For post-doctoral work, he entered research on ribosome function, the main emphasis being the introduction of fluorescence and rapid kinetic techniques for studying structure and dynamics of ribosome–ligand complexes. He was Associate Professor at the University of Munich from 1985, and became Professor of Molecular Biology at the University of Witten/Herdecke, Germany, in 1987. His present research interests include functional studies on ribsomes and translational elongation factors, in particular on the mechanism of translation, and on the bacterial signal-recognition peptide. **Marina V. Rodnina** studied Biology at the University of Kiev, Ukraine. For her PhD thesis, which she received in 1989 from the University of Kiev, she worked on the interaction of tRNA with eukaryotic ribosomes. From 1990 to 1992 she held a Postdoctoral Research Fellowship from the Alexander von Humboldt Foundation at the University of Witten/Herdecke. During that time she characterized two non-canonical binding sites for tRNA on rabbit liver ribosomes. Subsequently, at the same university, she moved into detailed kinetic studies on translational elongation-factor function in bacteria. She was appointed Associate Professor at the University of Witten/Herdecke in 1997. Her present research interests are mechanistic aspects of ribosome accuracy, GTPase mechanisms of elongation factors and the mechanism of translocation.

Mark Szczelkun obtained his BSc from the University of Liverpool in 1990 and his PhD from the University of Southampton in 1994. He currently holds a Wellcome Trust Career Development Fellowship at the University of Bristol. He has published papers on protein–DNA interactions, particularly communications between distant DNA sites mediated by restriction endonucleases and site-specific recombinases. Currently he is analysing the motor-protein activity of the type-I and type-III restriction endonucleases using a combination of biochemical and single-molecule techniques. Further information can be obtained at http://www.bch.bris.ac.uk/staff/mds.html.

Fergus D. Davison graduated from Otago University, New Zealand, and then pursued his scientific career in the UK after obtaining a PhD at the University of London. He is currently Senior Scientist in the Department of Cardiological Sciences at the same university, appointed in 1997. He is responsible for genetic analysis of and mutation detection in those patients presenting at the department with hypertrophic cardiomyopathy. **Leon G. D'Cruz** completed his MPhil in Immunology at the University of Essex in 1994 and spent a further 3 years at the MRC Centre, Cambridge, on a project developing immune-complement-based therapies. He is currently researching for his PhD on the molecular mechanisms for heart disease in hypertrophic cardiomyopathy. **William J. McKenna** graduated from Yale University and McGill University Medical School, Montreal, Canada, and pursued a career in cardiology at the Royal Postgraduate Medical School in London. In 1988 he moved to St. George's Hospital and is currently Professor of Cardiac Medicine. He is particularly interested in the diagnosis and management of patients with cardiomyopathy, sudden death in the young, sports cardiology and exercise physiology in health and disease. He is currently a council member of the International Society and Federation of Cardiology, patron of the charity Cardiac Risk in the Young and vice-president of the Cardiomyopathy Association.

Richard Libby received his BSc in Biology from Villanova University in 1990 and a PhD in Biology from Boston College in 1997. His doctoral dissertation concerned the roles of extracellular-matrix components in retinal development and function. Presently he is a postdoctoral fellow in Professor Karen Steel's Hereditary Deafness group at the MRC Institute of Hearing Research in Nottingham, UK. **Karen Steel** received a BSc in Genetics from the University of Leeds in 1974 and a PhD in Genetics from the Department of Human Genetics and Biometry at University College London in 1978. Her doctoral dissertation was entitled *Studies on Mice with Genetical and Experimentally Induced Abnormalities of the Inner Ear*. After her doctoral work she was a postdoctoral fellow at the Institute für Zoologie, Technische Universität München, and then joined the staff of the MRC Institute of Hearing Research in Nottingham. Over the past 17 years she has occupied several positions at this institute and is currently Senior Scientist and group leader of the Hereditary Deafness group and Special Professor of Genetics at the University of Nottingham. Over the last 20 years Professor Steel has written over 100 scientific publications concerning hearing and deafness. As a result of this work she won the Kresge-Mirmelstein Prize for excellence in hearing research in 1998. Presently she is serving on several editorial boards for scientific journals, is a member of the International Mouse Genome Nomenclature committee, and is a consultant to many scientific and charitable organizations that are involved with mouse genetics, and hearing and deafness.

Abbreviations

aa-tRNA	amino acyl-tRNA
AFM	atomic force microscope
BFP	blue fluorescent protein
CP	central pair
cTn	cardiac Tn
DHC	dynein heavy chain
DRC	dynein regulatory complex
E site	exit site
EF-2	elongation factor 2
EF-G	elongation factor G
EF-Tu	elongation factor Tu
EM	electron microscopic
FRET	fluorescence resonance energy transfer
GFP	green fluorescent protein
HCM	hypertrophic cardiomyopathy
HMM	heavy meromyosin
IC	intermediate chain
IDA	inner dynein arm
IFT	intraflagellar transport
KHC	kinesin heavy chain
LC	light chain
MT	microtubule
MTOC	microtubule-organizing centre
MyTH4	myosin tail homology 4
ODA	outer dynein arm
RS	radial spoke
S1	subfragment 1
S2	subfragment 2
SH3 domain	Src homology 3 domain
TIRF	total internal-reflection fluorescence microscopy
Tn C	troponin C
Tn I	troponin I
Tn T	troponin T

Introduction: the mechano-biochemistry of molecular motors

Rob Cross

Molecular Motors Group, Marie Curie Research Institute, Oxted, Surrey RH8 OTE, U.K.

Molecular motors are mechano-enzymes, protein machines for transducing chemical potential energy into mechanical motion. Some general principles of their mechanisms are known from now-classical biochemical and physiological studies on muscle. The catalytic cleavage of a nucleotide (usually ATP) in the active site drives the motor into a metastable, mechanically strained conformation, and the subsequent relaxation of this conformation to an unstrained state is harnessed to do useful work. To gain more detailed information about structure–function relationships we need to tinker with each motor, much as a motor mechanic tinkers with a macroscopic engine. The questions are the same. What moves? Why does it move? What limits its performance? What happens when I pull this?

Like the motor in your car, molecular motors have an operational cycle of forceful shape changes, and again like your car, this cycle is co-ordinated and driven by the chemistry of fuel consumption. For molecular-scale motors, it is the chemical catalytic cycle of ATP turnover in the motor active site that drives the development of mechanical strain. And just as the chemical transitions of ATP processing generate motion, so mechanical strain applied to the motor by an external load can feed back on the chemistry of the motor active site. This feedback allows the motor not only to generate force, but also to collaborate efficiently with other members of a team of motors. By sensing the forces produced by its team mates and adjusting its own force-generating schedule to fit in, a particular motor can contribute efficiently to a team effort. The problem, which we have with each molecular motor, is to understand the system of mechanical linkages, levers, springs, pressure plates, push-rods, bearings and even axles that amplify slight motions in the catalytic residues into larger-scale motions of the full molecule. The study of this reciprocal mechanochemical coupling is the linking theme of the essays in this volume.

Most of the setting up and maintenance of the large-scale organization in living cells relies on an interplay between motor-driven transport of components along cytoskeletal tracks (actin filaments and microtubules), and the (partly motor-dependent) continuous re-organization of cytoskeletal track. The classical motor is myosin II, the engine for skeletal-muscle contraction (Bagshaw) and also for smooth-muscle and cardiac-muscle contraction (Davison, D'Cruz and McKenna). More recent studies have revealed a plethora of previously unknown isoforms of myosin (Kalhammer and Bahler), kinesin (Kull) and dynein (Harrison and King), each specialized for a particular mechanical task. Critically, motor-dependent processes include mitosis and cytokinesis, the morphogenesis of the endoplasmic reticulum, the directional transport of Golgi, lysosomes and chloroplasts; exo-, endo- and pinocytosis, the polarization of the egg, gastrulation, cell crawling, muscle contraction (Bagshaw), the beating of cilia and flagella (Woolley), the actuation of hearing (Libby and Steel), and even thought itself [neuronal growth-cone motility and the axonal transport of neuortransmitter-containing vesicles (Signor and Scholey)]. Together, the track-following molecular motors kinesin, myosin and dynein are responsible not only for the transport and targetting of cellular components, but also for tensioning, sliding and rearranging cellular arrays of microtubules and actin filaments, and even governing their rates of assembly and disassembly. With only a modest leap of the imagination, even the bacterial rotary motors and and the bacterial F_1-ATPase (Kinosita, Yasuda and Noji) can be regarded as following an endless, circular track. The problem of the stepping mechanism by which the motor moves along its track is thus in common between muscle and non-muscle myosins, kinesin, dyneins, DNA-track followers (Szczelkun) and the GTPase mRNA-track follower elongation factor G (Wintermeyer and Rodnina). Solving this central problem is a demanding task, but the field has an enviable record of experimental ingenuity, including the development of entirely new types of single-molecule microscope (Knight and Molloy; Kinosita et al.), and there is every prospect of success. The scientific study of the best-understood motor, myosin, can be traced back 400 years to the Renaissance, but far from senescing, our field has not yet even reached maturity. What we are set to learn will have wide implications for biology, where the mechanical aspect of biochemical reactions is underappreciated, and, beyond that, for nanotechnology and molecular-scale robotics.

F$_\mathrm{I}$-ATPase: a highly efficient rotary ATP machine

Kazuhiko Kinosita, Jr.*†[1], Ryohei Yasuda† & Hiroyuki Noji†

Department of Physics, Faculty of Science and Technology, Keio University, Hiyoshi, Kohoku-ku, Yokohama 223-8522, Japan, and †CREST "Genetic Programming" Team 13, Teikyo University Biotechnology Center 3F, Nogawa, Miyamae-ku, Kawasaki 216-0001, Japan

Introduction

Think of a single protein molecule that is by itself a rotary motor. Driven by three subunits each fuelled by ATP, the motor rotates in discrete 120° steps. The efficiency of energy conversion, from the free energy of ATP hydrolysis to mechanical output of the motor, amounts to nearly 100%. Mother Nature has created such a tiny yet powerful molecular machine, not for the purpose of producing mechanical work but for the synthesis of ATP in our body by reverse operation of the rotary motor.

This rotary motor is a part of the enzyme ATP synthase. In animals, the ATP synthase resides in the inner membrane of mitochondria. The food ingested by an animal is 'burnt' (oxidized) by protein machinery embedded in the inner membrane, and the energy obtained by the oxidation is used to eject protons from inside the mitochondrion to the external space. The protons eventually flow back into mitochondria through the ATP synthase, in which

[1]*To whom correspondence should be addressed, at Keio University.*

ATP is synthesized from ADP and P$_i$ using the proton flow as the energy source. Similar systems occur in plants and bacteria.

That oxidation and ATP synthesis are coupled by the flow of protons across the mitochondrial membrane was proposed by Peter Mitchell [1], a radical concept at that time which took many years to be accepted. For the coupling between the proton flow and ATP synthesis in the ATP synthase, another revolutionary proposal was made by Boyer [2]: the proton flow and chemical reaction are coupled by the mechanical rotation of a subunit(s) within the protein molecule. This latter proposal, too, failed to arouse enthusiasm until, in 1994, John Walker and colleagues elucidated the atomic structure of part of the ATP synthase [3]. The structure strongly supported Boyer's idea, and also suggested many experiments that have led to the proof (at least in part) of Boyer's rotational catalysis model [4–8]. Here we briefly review some of the remarkable features of this molecular machine revealed in our laboratory, and discuss its possible mechanism.

The ATP synthase: two rotary motors with a common shaft

As shown in Figure 1(a), the ATP synthase consists of two parts, a membrane-embedded portion called F$_0$ and a protruding portion F$_1$. When protons flow through F$_0$ from top to bottom in Figure 1(a), ATP is synthesized in F$_1$. The ATP synthase is a completely reversible machine: when ATP is hydrolysed in F$_1$, protons are pumped back in the reverse direction.

Boyer [2,9,10] proposed that F$_0$ is a rotary motor (or rather a turbine) driven by the proton flow, and that F$_1$ is another rotary motor driven by ATP hydrolysis. The two motors have a common rotary shaft (magenta in Figure 1a), but the genuine rotary directions of the two are different. When the free energy liberated by the downward flow of protons is greater than the free energy of ATP hydrolysis, the F$_0$ motor rotates the common shaft in the F$_0$'s genuine direction. The F$_1$ motor is forcibly rotated in its reverse direction, resulting in ATP production in its catalytic sites. If the free energy of ATP hydrolysis is higher, the F$_1$ motor gains control and rotates the shaft in its own direction. Protons are then pumped out by F$_0$ against an uphill potential.

Isolated F$_1$ catalyses only ATP hydrolysis, and hence is called F$_1$-ATPase. Its subunit composition is $\alpha_3\beta_3\gamma\delta\epsilon$. One view of the crystal structure of bovine mitochondrial F$_1$, determined by Walker and colleagues [3], and hereafter referred to as the Walker structure, is shown in Figure 1(b). The δ and ϵ subunits were not resolved, but these are not required for the rotation of F$_1$. The $\alpha_3\beta_3$ cylinder forms the stator, and the central γ subunit, of which part of the protruding portion has not been resolved, would rotate in the cylinder. The catalytic sites in which ATP is synthesized/hydrolysed are on the three β subunits, each at an interface with a neighbouring α. Surprisingly, each of the three β subunits carried a different nucleotide in the crystal: one an analogue of ATP, another ADP, and the third carried none, in the clockwise order

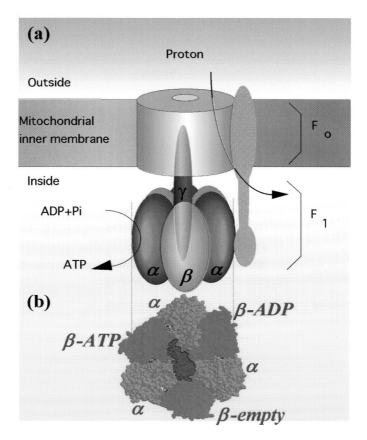

Figure 1. ATP synthase
(a) A schematic model. In a currently popular but unproven view, the red cylinder in the membrane rotates together with the γ shaft, and the grey part serves as the stator. Adapted from [12] with permission. ©1998 Cell Press. (b) Top view of the atomic structure of the F_1 part [3].

(Figure 1b). If hydrolysis were to proceed from this crystal structure, the ATP in the first β would be hydrolysed into ADP, the ADP in the second β would be released, and the third β would bind ATP from the medium. Thus the central γ is expected to rotate anticlockwise.

Proof that F_1 is indeed a rotary motor

Large-amplitude rotational motion of γ during ATP hydrolysis or synthesis had been shown by crosslinking and spectroscopic studies [4–7], but whether γ makes complete turns and does so in a unique direction was not clear until we observed the motion of γ directly under a microscope [8]. To visualize the rotation, Hiroyuki Noji prepared an $\alpha_3\beta_3\gamma$ subcomplex of bacterial origin. The $\alpha_3\beta_3$ cylinder was fixed to a glass surface, and a micrometre-sized actin filament was attached to γ via streptavidin (Figure 2a). The actin filament was

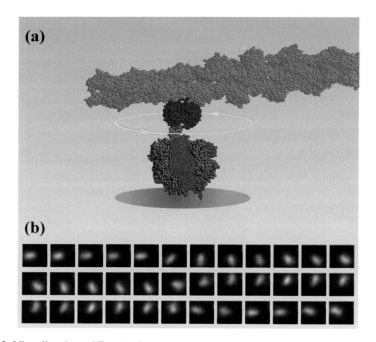

Figure 2. Visualization of F₁ rotation

(a) Orange rod at the top, actin filament; dark brown, streptavidin; magenta, γ subunit of F_1; green, β subunit; blue, α subunit. The diameter of the grey disk is $\approx$22 nm. (b) Photographs at 26 ms intervals of the actin filament rotating stepwise at 0.6 μM ATP.

fluorescently labelled. When Ryohei Yasuda looked into a fluorescence microscope, he saw a filament rotating continuously in one direction, anticlockwise as expected! The rotating filament was found on the very first day, within 30 min of the initial trial. The rotation was so beautiful that we were immediately convinced, almost, that F_1 was indeed a rotary motor. Soon a second rotating filament was found, and we were to toast Boyer after confirming the third. But the beer remained unopened; the third came only after a full month of struggle.

In our hands, at most a few percent of the actin filaments in a sample chamber rotate. A likely cause is surface obstructions. Note that the F_1 molecule is only $\approx$10 nm high, whereas the actin filaments are $\approx$1 μm or longer. Thus rotating the filament without touching the surface should be difficult. Indeed, rotating filaments often show a tendency to be stuck at a particular angle. Also, a significant fraction of F_1 is idle under normal assay conditions: MgADP, a product of the ATPase reaction, tends to bind tightly to a catalytic site and inhibit further ATP hydrolysis. Presumably, this MgADP inhibition prevents futile consumption of ATP in living cells.

Figure 3 shows typical (i.e. the most vigorously rotating) examples of rotation versus time. At a saturating concentration of ATP (2 mM), the rotation was essentially smooth and unidirectional. Rotation was slower for longer

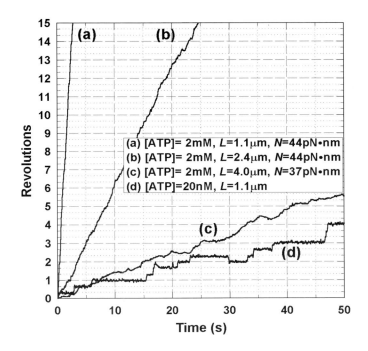

Figure 3. Rotation at high and low ATP concentrations
Traces (a–c) show the rotational rates for actin filaments of different lengths at 2 mM ATP.
Analysis of the data [13] gives torque values of 44, 44 and 37 pN·nm, respectively. The rotational
rates during individual steps at 20 nM ATP (trace d) are similar to the rate for the same filament
length at 2 mM ATP (trace a).

actin filaments, because the viscous friction against the rotation is basically
proportional to the cube of the actin length [11]. At very low ATP concentrations, the rotation became stepwise (Figures 2b and 3, trace d). The step size of
120° is precisely the one expected for the motor driven by the three β subunits
separated by 120°. Note, in Figure 3 (trace d), that the motor made a back step
at ≈30 s; a molecular machine must occasionally make mistakes.

The properties of the F$_1$ motor

Our studies [8,12–14] have revealed the following properties of F$_1$-ATPase.
These properties have been deduced from the observations of the most actively
rotating F$_1$, and do not necessarily represent the average behaviour of an
ensemble.

(i) F$_1$-ATPase is a rotary motor made of a single molecule
Occasionally an actin filament rotated around its centre, like a propeller [8]. If
one were to hold a long rod at the middle and rotate it like a propeller, one
would have to shift one's grip continually; true rotation requires slippage
between the rotor and stator, compared with the pseudo-rotation that one can

make by holding the end of a rod and twisting (not really rotating) the wrist. Thus the γ subunit must slide against the surrounding $\alpha_3\beta_3$ cylinder over infinite angles. The propeller rotation of an actin filament cannot be supported by two different F_1 molecules, and therefore a single F_1 molecule must itself be a rotary motor. Its diameter and height being only ≈ 10 nm [3], the F_1 motor is the smallest rotary motor known.

(ii) $\alpha_3\beta_3\gamma$ subunits suffice for rotation

We have demonstrated rotation in the subcomplex $\alpha_3\beta_3\gamma$ [8]. Crosslinking studies [7] have indicated that the ϵ subunit also moves relative to α, and rotation of an actin filament attached to ϵ has been demonstrated [14]. Thus ϵ is likely to be part of the rotor, although it is not a necessary part of the rotary mechanism.

(iii) Rotation is anticlockwise when viewed from the F_0 side

Except for the occasional back steps, the sense of rotation is always anticlockwise when viewed from the top in Figures 1(a) and 2(a) [8]. This direction is in accord with the Walker structure (Figure 1b), suggesting that a structure similar to that in Figure 1(b) appears during rotation.

(iv) The F_1 motor is a 120° stepper

Stepwise rotation is seen at submicromolar ATP concentrations [13] (Figures 2b and 3, trace d). Between steps, F_1 waits for the next ATP molecule to arrive. At high ATP concentrations, the waiting time is shorter than the time required to rotate an actin filament through 120°, and hence the steps are not easily discerned. So far, substeps within the 120° step have not been resolved at our highest temporal resolution of 5 ms.

(v) The F_1 motor is designed to produce a constant torque

From the measured rate of rotation, we can calculate the torque the F_1 motor produces to move the actin filament [13]. At saturating ATP concentrations where ATP binding is not rate limiting, the torque, averaged over many revolutions, is ≈ 40 pN·nm irrespective of the viscous load or the rotational rate [13] (see also Figure 3). At low ATP concentrations where the 120° steps are resolved, the torque driving each step averages ≈ 44 pN·nm, again irrespective of the filament length [13] (see also Figure 4).

(vi) Work per step is also constant

Mechanical work done (against the viscous load) in a single step is given by the angular displacement, $2\pi/3$ radians (i.e. 120°), multiplied by the torque. Because the torque is constant, 40–44 pN·nm, the work done in a step is also constant and amounts to 80–90 pN·nm.

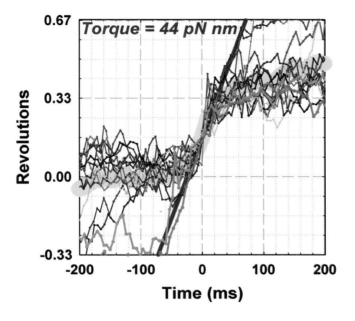

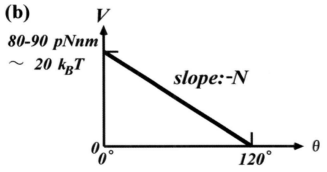

Figure 4. Time courses of individual steps
(a) Steps were measured at 0.2 μM ATP for F_1 bearing a 1.0 μm filament. Traces for individual steps are superimposed such that the middle of each step is located at time zero. Rapid succession of two steps are seen in two traces (red and pink). A thick cyan line shows the average of all traces. A thick green line shows a constant-speed rotation at 44 radians/s, corresponding to a torque of 44 pN·nm. (b) Rotational potential, $V(\theta)$, deduced from (a).

(vii) One ATP molecule is consumed per step

At low ATP concentrations where ATP binding is rate-limiting both for hydrolysis and rotation, the time-averaged rate of actin rotation was approximately equal to one-third of the number of ATP molecules hydrolysed per s [13]. This suggests that one ATP molecule is consumed per 120° step. Both the rate of rotation and the rate of ATP hydrolysis were proportional to the ATP concentration, indicating that neither of these processes required simultaneous consumption of two or more ATP molecules. At very low ATP concentrations where the 120° steps were resolved, an analysis of the intervals between steps indicated that each step was fuelled by one ATP molecule [13].

(viii) The efficiency of energy conversion can reach ≈100%

Points (vi) and (vii) above indicate that the constant mechanical output of 80–90 pN·nm per step is produced by the consumption of one ATP molecule. The energy input, the free energy of ATP hydrolysis, ΔG, depends on the concentrations of ATP, ADP and P_i. In experiments at a controlled ΔG of 90 pN·nm, the work done per step was also ≈80 pN·nm [13]. Thus the F_1 motor can work at near 100% efficiency. The efficiency is lower at higher ΔG, because the mechanical output is constant. In cells where ΔG is 80–90 pN·nm, the efficiency could be ≈100%. This amazingly high efficiency, compared with other molecular motors [12], is probably related to the fully reversible nature of this molecular machine. In ATP synthase, the F_1 motor would pump protons at the energy conversion efficiency of ≈100%.

The rate of ATP hydrolysis quoted in point (vii) was measured in solution, and thus is an ensemble average. The hydrolysis rate of active F_1 might be higher, because some in the ensemble might have been inhibited. Hence, we cannot exclude the possibility that uncoupled, futile consumption of ATP occurs occasionally. Even so, each mechanical step is coupled to the hydrolysis of one and only one ATP molecule, and the free energy liberated in the coupled hydrolysis can be converted to mechanical work at ≈100% efficiency.

(ix) Bi-site catalysis supports rotation

At submicromolar ATP concentrations, the F_1-ATPase operates in the so-called bi-site mode [9,10], where at most two catalytic sites are filled with a nucleotide. Basically, one site binds tightly a nucleotide, ATP or ADP+P_i in reversible equilibrium, and the other two sites are empty. When a second site binds ATP from the medium, ADP and P_i are rapidly released from the first site, resulting in net hydrolysis of one ATP molecule. Our results show that rotation can occur in this bi-site catalysis: there is no need to fill all the catalytic sites. The bi-site rotation is fundamental to the rotary mechanism, and there is no clear evidence that the motor adopts a different mechanism in the tri-site regime. Because torque generation requires broken symmetry, the use of an all-filled (or all-empty) state is not advantageous.

At extremely low ATP concentrations, uni-site catalysis occurs where the tightly bound products in one catalytic site are very slowly released in the medium without the binding of a second ATP. Whether the uni-site catalysis accompanies rotation is an important yet unsettled question. Crosslinking γ to a β did not inhibit uni-site catalysis [15], indicating that uni-site catalysis can occur without rotation. However, the possibility of rotation in the absence of crosslinking cannot be dismissed.

(x) Back steps occur, probably using ATP

Occasional back steps observed at low ATP concentrations were as rapid as the forward steps [13]. The torque driving the back steps, and thus the work per step, are as high as those of the forward steps, suggesting that the back steps are also driven by ATP hydrolysis. Presumably, ATP binding to the wrong site (one of the two empty sites) in the bi-site rotation produces a back step.

(xi) ATP hydrolysis is likely to introduce a linear downhill rotational potential

Point (vi) can be explained if an angle-dependent potential energy of height 80–90 pN·nm, downhill towards the position 120° ahead, is introduced for γ rotation upon the binding (and/or subsequent hydrolysis) of ATP. We can estimate the shape of this potential as follows.

In Figure 4(a), many steps in a rotation record are superimposed. Although individual traces are noisy, their average shown in the thick cyan line indicates that the slope, the rotational rate ω, is approximately constant throughout the 120° interval. The torque N ($=\omega\xi$; where ξ is the rotational frictional drag coefficient [11,13], $\xi=1.0$ pN·nm·s for the 1 μm filament) is thus independent of the rotational angle θ, and is approximately 44 pN·nm throughout the 120° interval as shown in the thick green line. Because the potential $V(\theta)$ for γ rotation is related to N by $dV/d\theta=-N$, a linear potential profile is deduced, shown in Figure 4(b). Note that the actual potential profile should be dependent on the state of the bound nucleotides, which changes with time in each step. The profile shown in Figure 4(b) is the effective potential experienced by γ during the course of the chemical kinetics. Also, details of the actual potential profile may have been smoothed out by the combination of possible elastic linkage between γ and actin and the viscous friction on the latter.

How the F₁ motor may be designed

A rotary mechanism of human design is shown in Figure 5. This motor has three driving poles in the stator part, like the F_1 motor. The motor is powered by a unidirectional current source, while ATP hydrolysis is also practically unidirectional. Thanks to the three pairs of switches (commutators) on the shaft, the three poles change their polarities such that the shaft rotates continuously in the anticlockwise direction. The rotor is a permanent magnet, a static component. If the rotor is forcibly rotated in the reverse direction, this DC (direct current) motor becomes a DC generator and charges the external battery. The energy-conversion efficiency of modern electrical motors is quite high, often >95%. The DC motor operates in the bi-site mode in that the three driving poles never assume the same polarity. Thus there are similarities

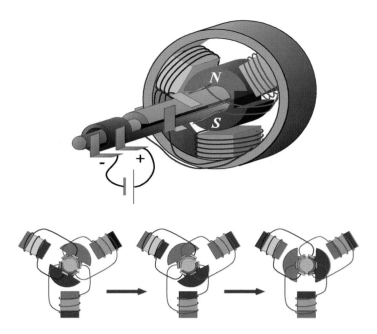

Figure 5. A three-pole DC motor
The commutators on the shaft change the polarity of stator magnets such that the shaft rotates anticlockwise continuously.

between the properties of this three-pole DC motor and those of the F_1 motor. Do these two share some operational principles?

The driving forces in the electrical motor in Figure 5 are the attraction between north and south poles and the repulsion between like poles. Such a push–pull mechanism may also operate between the γ and β subunits of the F_1 motor. Figure 6(a) shows side views of the three pairs of opposing β and α subunits, together with the central γ, in the Walker structure. The vertical black lines show the rotational axis suggested by Wang and Oster [16]: the bottom part of the $\alpha_3\beta_3$ stator has an approximate 3-fold symmetry around this axis, and thus the conformations of β and α in the bottom do not change greatly depending on the bound nucleotide. In the upper part, in contrast, the β subunits binding ATP or ADP are bent towards, and therefore push, γ, whereas the empty β retracts and pulls γ towards it. Wang and Oster [16] suggest that, because the central γ is slightly bent, co-operative push–pull actions of the three β subunits would rotate γ, as seen in Figure 6(b).

A simple F_1 model

Figure 7 shows a model for F_1 rotation based on this push–pull mechanism. The side of γ that faces the empty β in the Walker structure is designated the north pole, and thus an empty β is the south pole. A nucleotide-carrying β is north and repels the north face of γ and attracts its south face. By reciprocity, the south face of γ augments the affinity of the opposing β for a nucleotide,

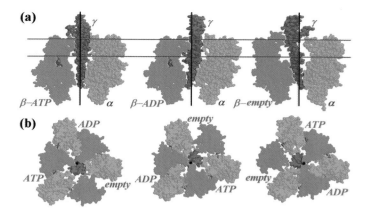

Figure 6. Nucleotide-dependent conformational changes in F₁

(a) The diagrams show the central γ subunit (orange), one β (green) to the left of γ, and one α (blue) to the right in the crystal structure [3]. The black lines indicate the rotation axis [16]. Nucleotides are shown in Corey-Pauling-Kultun (CPK) colours. (b) Top view of the cross sections of F₁ between the horizontal lines in (a).

and the north face decreases the affinity (the free energy is lowered when north and south oppose each other). To ensure rotation in a unique direction, additional control of nucleotide-binding kinetics via 'commutators' is required. A simple example is given in Figure 7(b): binding/release of ATP is allowed for β while it is on the pink side of γ, and ADP binding/release while on the green side. As shown in Figures 7(c) and 7(d), the switching ensures anticlockwise rotation of γ when ATP is hydrolysed; when the rotor is forced to rotate clockwise in the presence of ADP (and P$_i$), ATP is synthesized.

More elaborate switches have been proposed by Wang and Oster [16], and their model can account for many experimental observations, including the near 100% efficiency and 120° stepping with occasional back steps. In their model, as with the model in Figure 7, the motor tends to pause at angles 60° out of phase from the Walker structure (Figure 7a), an experimentally testable prediction. How the switching action is implemented in the protein structure is yet to be specified.

A switch-less F₁ model

The roles of the commutators in Figure 5 are to alternate the polarities of the stator magnets, and to do so at precise timings dictated by the rotational angle of the shaft. Unlike the magnet driven by direct current, the alternation of the polarity is inherent in the ATP-driven 'magnet', where bound ATP is eventually hydrolysed and released, restoring the south state spontaneously. Thus only co-ordination of nucleotide kinetics among the three stator magnets needs to be programmed. This could be done without switches, as shown in Figure 8, which is one version of the general model of Oosawa and Hayashi [17].

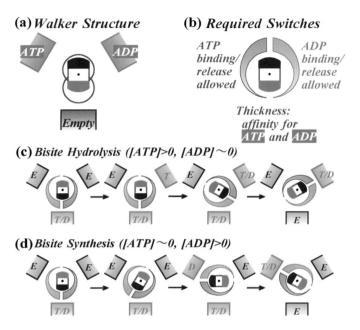

Figure 7. A simple model for the F$_1$ motor
(a) The γ subunit is regarded as a permanent magnet, the side of γ that faces the empty β in the Walker structure being the north pole. (b) The affinity for a nucleotide is higher when β is closer to the south pole of γ (strictly, the affinity for ATP is higher than that for ADP). Binding and release of ATP and ADP are kinetically inhibited on the green and pink sides, respectively. (c) Bound ATP (T) is in equilibrium with ADP (D) and P$_i$, and is released as ADP when a second ATP binds and rotates γ. E, empty. (d) Forced clockwise rotation of γ results in the uptake of ADP (and P$_i$) and release of ATP.

In Figure 8, the position of the 'magnetic pole' in β changes depending on the bound nucleotide. This dual-pole arrangement, combined with the higher affinity for nucleotides when the pole is closer to the south face of γ, ensures anticlockwise rotation in bi-site hydrolysis by inducing ATP binding primarily in the empty β in the anticlockwise direction (Figure 8c). Note that the angle-dependence of the nucleotide affinity results as a reaction to the nucleotide-dependent push–pull action, without requiring switches. An additional factor ensuring correct rotation is the higher affinity for ATP than for ADP. Because ATP hydrolysis on β is reversible (the free-energy difference between β binding ATP and β binding ADP+P$_i$ is small), the affinity for the hydrolysis products has to be lower in order for β to act as an ATPase. In the original model of Oosawa and Hayashi [17], near-100% efficiency was achieved for both hydrolysis and synthesis.

The essence of the dual-pole arrangement is that the mechanical interaction between γ and individual β subunits involves a nucleotide-dependent rotational component in addition to pushing/pulling. The Walker structure gives more emphasis to pushing/pulling rather than to direct rotation, but the

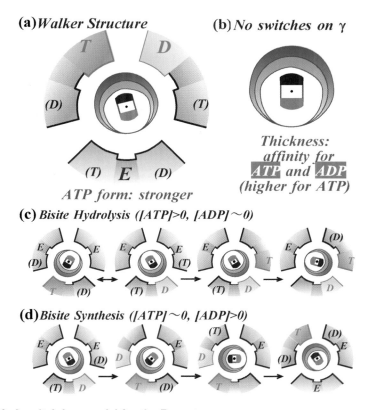

Figure 8. A switch-less model for the F$_1$ motor
(a) Location of the magnetic pole on β changes depending on the bound nucleotide. ATP (T) magnet is stronger than ADP (D) magnet and, hence, (b) the affinity for ATP is higher than that for ADP. (c) When only one nucleotide is bound it is reversibly interconverted between ATP and ADP+P$_i$. Comparison of affinities suggests that the most likely way of filling a second site is binding of ATP in the β in the anticlockwise direction. (d) When γ is forcibly rotated clockwise, the equilibrium between ATP and ADP is shifted towards ATP, which is eventually released while ADP is newly bound in the second site. E, empty.

structure of γ in the protruding portion has not been resolved. Also, key intermediates in actual rotation, particularly the state with only one bound nucleotide, may well have a different structure.

Models in Figures 7 and 8 have been introduced solely to point out several factors that may or may not be important in the mechanism of F$_1$ rotation. Neither is assumed to be the actual mechanism. Nor are all important factors explained in these models. For example, the magnet analogy, particularly that for γ, obscures the fact that neither γ nor β possesses reflection symmetry. The force between γ and individual β subunits must be more or less asymmetric, favouring one rotational direction over the other, as modelled by the dual poles in Figure 8. Whether the rotational potential can be approximated by a simple superposition of three pairwise interactions between individual β and γ, as implied by the magnetic analogy, remains unclear until structures of the different intermediates are revealed. Also, in Figures 7 and 8, nucleotide kinetics on different βs are co-ordinated only through the rotation of γ, but β subunits may also communicate through the intervening α subunits.

Problems to be solved

The key to understanding the rotational mechanism is to elucidate the structure of F_1 in which only one catalytic site is filled, the structure from which the 120° step begins. It is unlikely that the two empty β subunits in this structure both resemble the empty β in the Walker structure; presumably, the asymmetric γ would induce the β 120° ahead into a conformation closer to the nucleotide-carrying β in the Walker structure.

Also important is to establish the precise relation between the nucleotide binding/hydrolysis kinetics and the rotational potential. This could be done by imaging nucleotide turnover in a single motor molecule [18] while observing its rotation through an attached actin filament. Manipulation of the filament, e.g. with optical tweezers [19], will help establish the angle dependence of the nucleotide kinetics or, conversely, the nucleotide dependence of the rotational torque. Because an attached actin filament may not faithfully reflect the orientation of γ, assessment of the latter through imaging of polarized fluorescence [20,21] from a fluorophore rigidly attached to γ will also be useful.

What happens if the free energy of ATP hydrolysis is reduced below 80 pN·nm by manipulating nucleotide and P_i concentrations? This is a fundamental question for mechanisms of molecular machines in general. The predicted behaviours depend on the model. Answering the question experimentally is not easy, because MgADP inhibition is serious at high ADP concentrations.

Presumably, the MgADP-inhibited form is the most stable state of the F_1 motor, while rotation requires instability. The Walker crystal structure probably represents this stable inhibited form [3]. The anticlockwise rotation consistent with this structure, then, implies that slight destabilization of the Walker structure, e.g. by the presence of P_i next to ADP, would make an active intermediate. Interestingly, the inhibition does not occur in the synthesis mode [10], where proton-driven rotation of γ may destabilize the inhibited form. Many articles on F_1-ATPase do not mention the degree of MgADP inhibition in the experiments described; in some cases, different interpretations could emerge if this almost inevitable inhibition was taken into account.

Will ATP be synthesized in F_1 without the aid of F_0, if one mechanically rotates γ clockwise, e.g. by manipulating attached actin with optical tweezers? The answer should be yes, but experimental proof is still awaited. Such an experiment would demonstrate that mechanical energy can be directly transformed into chemical energy.

Relatively little is known about the F_0 part of the ATP synthase. Even whether F_0 is really a rotary motor is yet to be proved. If it is, which part is the rotor and how is it connected to γ? Is the proton transport tightly coupled to the rotation as it seems to be between ATP hydrolysis and rotation in the F_1 motor? Many questions remain, demanding new experimental ideas. Boyer calls the ATP synthase a splendid molecular machine [10]. It is also a splendid toy for young, creative researchers.

Summary

- *A single molecule of F_1-ATPase is by itself a rotary motor in which a central subunit, γ, rotates against a surrounding stator cylinder made of $\alpha_3\beta_3$ hexamer.*
- *Driven by the three β subunits that hydrolyse ATP sequentially, the motor runs with discrete 120° steps at low ATP concentrations.*
- *Over broad ranges of load and speed, the motor produces a constant torque of 40 pN·nm.*
- *The mechanical work the motor does in the 120° step, or the work per ATP hydrolysed, is also constant and amounts to 80–90 pN·nm, which is close to the free energy of ATP hydrolysis. Thus this motor can work at near 100% efficiency.*

We are grateful to Professor M. Yoshida and the members of CREST (Core Research for Evolutionary Science and Technology) Team 13 for collaboration and discussion. This work was supported in part by Grants-in-Aid from the Ministry of Science, Education, Sports and Culture of Japan, and a Keio University Special Grant-in-Aid. R.Y. was a Research Fellow of the Japan Society for the Promotion of Science.

References

1. Mitchell, P. (1961) Coupling of phosphorylation to electron and hydrogen transfer by a chemi-osmotic type of mechanism. *Nature (London)* **191**, 144–148

2. Boyer, P.D. & Kohlbrenner, W.E. (1981) The present status of the binding-change mechanism and its relation to ATP formation by chloroplasts, in *Energy Coupling in Photosynthesis* (Selman, B.R. & Selman-Reimer, S., eds.), pp. 231–240, Elsevier, Amsterdam

3. Abrahams, J.P., Leslie, A.G.W., Lutter, R. & Walker, J.E. (1994) Structure at 2.8 Å of F_1-ATPase from bovine heart mitochondria. *Nature (London)* **370**, 621–628

4. Duncan, T.M., Bulygin, V.V., Zhou, Y., Hutcheon, M.L. & Cross, R.L. (1995) Rotation of subunits during catalysis by *Escherichia coli* F_1-ATPase. *Proc. Natl. Acad. Sci. U.S.A.* **92**, 10964–10968

5. Zhou, Y., Duncan, T.M., Bulygin, V.V., Hutcheon, M.L. & Cross, R.L. (1996) ATP hydrolysis by membrane-bound *Escherichia coli* F_0F_1 causes rotation of the γ subunit relative to the β subunits. *Biochim. Biophys. Acta* **1275**, 96–100

6. Sabbert, D., Engelbrecht, S. & Junge, W. (1996) Intersubunit rotation in active F-ATPase. *Nature (London)* **381**, 623–625

7. Aggeler, R., Ogilvie, I. & Capaldi, R.A. (1997) Rotation of a γ-ϵ subunit domain in the *Escherichia coli* F_1F_0-ATP synthase complex. *J. Biol. Chem.* **272**, 19621–19624

8. Noji, H., Yasuda, R., Yoshida, M. & Kinosita, Jr., K. (1997) Direct observation of the rotation of F$_1$-ATPase. *Nature (London)* **386**, 299–302

9. Boyer, P.D. (1993) The binding change mechanism for ATP synthase — some probabilities and possibilities. *Biochim. Biophys. Acta* **1140**, 215–250

10. Boyer, P.D. (1997) The ATP synthase - a splendid molecular machine. *Annu. Rev. Biochem.* **66**, 717–749

11. Hunt, A.J., Gittes, F. & Howard, J. (1994) The force exerted by a single kinesin molecule against a viscous load. *Biophys. J.* **67**, 766–781

12. Kinosita, Jr., K., Yasuda, R., Noji, H., Ishiwata, S. & Yoshida, M. (1998) F$_1$-ATPase: a rotary motor made of a single molecule. *Cell* **93**, 21–24

13. Yasuda, R., Noji, H., Kinosita, Jr., K. & Yoshida, M. (1998) F$_1$-ATPase is a highly efficient molecular motor that rotates with discrete 120° steps. *Cell* **93**, 1117–1124

14. Kato-Yamada, Y., Noji, H., Yasuda, R., Kinosita, Jr., K. & Yoshida, M. (1998) Direct observation of the rotation of γ subunit in F$_1$-ATPase. *J. Biol. Chem.* **273**, 19375–19377

15. García, J. J. & Capaldi, R. A. (1998) Unisite catalysis without rotation of the γ-ϵ domain in *Escherichia coli* F$_1$-ATPase. *J. Biol. Chem.* **273**, 15940–15945

16. Wang, H. & Oster, G. (1998) Energy transduction in the F$_1$ motor of ATP synthase. *Nature (London)* **396**, 279–282

17. Oosawa, F. & Hayashi, S. (1986) The loose coupling mechanism in molecular machines of living cells. *Adv. Biophys.* **22**, 151–183

18. Funatsu, T., Harada, Y., Tokunaga, M., Saito, K. & Yanagida, T. (1995) Imaging of single fluorescent molecules and individual ATP turnovers by single myosin molecules in aqueous solution. *Nature (London)* **374**, 555–559

19. Arai, Y., Yasuda, R., Akashi, K., Harada, Y., Miyata, H., Kinosita, Jr., K. & Itoh, H. (1999) Tying a molecular knot with optical tweezers. *Nature (London)* **399**, 446–448

20. Sase, I., Miyata, H., Ishiwata, S. & Kinosita, Jr., K. (1997) Axial rotation of sliding actin filaments revealed by single-fluorophore imaging. *Proc. Natl. Acad. Sci. U.S.A.* **94**, 5646–5650

21. Hälser, K., Engelbrecht, S. & Junge, W. (1998) Three-stepped rotation of subunits γ and ϵ in single molecules of F-ATPase as revealed by polarized, confocal fluorometry. *FEBS Lett.* **426**, 301–304

Motors in muscle: the function of conventional myosin II

Clive R. Bagshaw

Department of Biochemistry, University of Leicester, Leicester LE1 7RH, U.K.

Introduction

The study of the mechanism of muscle contraction has a long history and the ideas that have emanated from this field have laid the foundation for much of what we know about molecular motors. Skeletal muscle, in particular, has been the target for investigation because its structural organization provides clues to the mechanism of contraction, and the abundance of contractile proteins within it allows ready isolation of its components for biochemical characterization [1,2]. In skeletal muscle, the key proteins, actin and myosin (of the conventional two-headed myosin-II class) are organized in linear arrays, making up the so-called thin and thick filaments of the sarcomere (Figure 1). A key observation in the 1950s was that, during contraction, these filamentous assemblies did not shorten but slid past one another.

Two decades earlier, myosin was shown to have ATPase activity; a crucial result at a time when ATP was identified as a central player in bioenergetics. However, demonstrating that ATP is the prime energy source for muscle contraction proved difficult, owing to the efficient regenerating systems *in vivo*. The advent of motility assays *in vitro* in the 1980s marked a climax in the biochemical approach to understanding the molecular basis of contraction; protein-filament sliding could be observed by eye in real time using an optical microscope. A minimal system comprising purified F-actin, the myosin head domain and MgATP was demonstrated to be sufficient for mechanochemical coupling. The intricate organization of skeletal muscle is an adaptation for achieving rapid macroscopic movement and to develop high tension, but the fundamental transduction event is a property of a unitary interaction between two proteins.

Over these last two decades there have been impressive advances in technology that have opened up the field of molecular motors beyond the specialized topic of muscle contraction. Use of recombinant proteins and sensitive functional assays has removed some of the early advantages of investigating muscle proteins. Nevertheless, muscle remains a particularly attractive system for investigation because the conclusions from the study of unitary events at the protein level can be critically evaluated at the macroscopic level. Only when agreement is reached can we claim to have a clear understanding of the processes involved.

Identification of myosin crossbridges

Following the demonstration that muscle shortening involved the interdigitation of thick and thin filaments, the problem to be solved was rephrased as 'what makes filaments slide?'. On the basis of a number of properties of intact muscle, including that the tension which developed was proportional to the degree of overlap of the filaments whereas shortening velocity was independent, A.F. Huxley proposed that sliding was driven by the repetitive action of individual 'side pieces' that transiently connected to two filament types. In the same year, H.E. Huxley showed, using electron microscopy, the existence of protrusions from the thick filament that could link to the actin filament. The concept of the myosin crossbridge was born (for reviews, see [1,2]).

Solubilization of the thick filament yielded myosin as the dominant component; a protein with two pear-shaped heads about 16 nm long attached to a 160-nm-long tail. At physiological ionic strength, the tails pack together to form the filament backbone, while the heads project to make up the crossbridges. The tail is known to be a coiled-coil α-helix from characteristic X-ray fibre diffraction spots, whereas the head structure was solved by crystallography [3]. This form of myosin was termed myosin II to distinguish it from the single-headed myosin-I class and other members of the myosin superfamily identified subsequently. Structural analysis of the myosin-II molecule (termed simply myosin in the remainder of this chapter) benefited from partial proteolysis studies. Digestion with a number of proteases of different specificities showed accessible regions near the base of each head, yielding individual subfragment 1 (S1) moieties, and also about one-third of the way into the tail, yielding a two-headed fragment containing a short tail (heavy meromyosin, HMM). The remainder of the tail (light meromyosin) forms filamentous structures at physiological ionic strength, whereas the short tail attached to HMM (subfragment 2, S2) is soluble at low ionic strength.

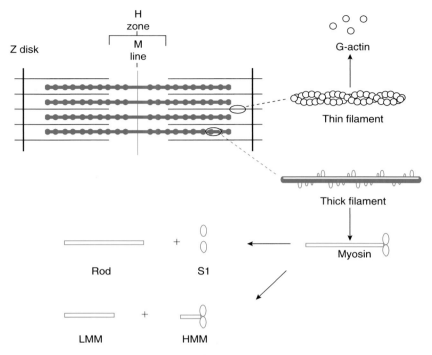

Figure 1. Schematic diagram of a sarcomere, the organelle of contraction in striated muscle

Disruption of the lattice allows the isolation of thick and thin filaments which can be solubilized to yield primarily myosin and actin, respectively. LMM, light meromyosin; HMM, heavy meromyosin.

The action of crossbridges has been studied from a number of different standpoints and using preparations ranging from intact muscle to isolated proteins. The regular arrangement of actin and myosin molecules within filaments and of the filaments within the muscle fibre has greatly aided the analysis by electron microscopy and X-ray diffraction. Early electron micrographs established that the myosin crossbridges attach to the actin filaments when the muscle fibre is depleted of ATP (i.e. the rigor state), in line with physiological measurements showing a very high stiffness. In the presence of ATP and the absence of Ca^{2+}, the crossbridges are detached and the muscle is flaccid (i.e. the relaxed state). On stimulation of intact muscle, or addition of ATP plus Ca^{2+} in the case of permeabilized preparations, tension is developed if the muscle is prevented from shortening (i.e. the isometric state), but analysis of the structural state(s) of the crossbridge by electron microscopy is difficult because the fixation procedures are likely to perturb the distribution of dynamic states. More recently, rapid freezing followed by cryoelectron microscopy has been used in an attempt to overcome this problem [4].

X-ray diffraction of intact muscle provides a complementary non-destructive approach. Several of the diffraction spots in the pattern can be assigned to particular spacings between protein units within the muscle and clear changes are seen on going from the relaxed to the contracted states and to the rigor state. The combined conclusions from these structural studies are that in rigor most, if not all, the myosin heads attach to the actin with a high degree of stereospecificity, with the main axis of the myosin head lying ≈45° to the fibre axis. In relaxed muscle most of the heads lie close to the filament backbone, while in isometric contraction there is more disorder, indicating that a distribution of crossbridge states exists, although a significant proportion appears to be attached.

The angled attached crossbridge in the rigor state is thought to represent the end-state of a contraction cycle. If the crossbridge was to attach to the actin filament at an angle near to the normal and then tilt to the rigor position, a relative translational movement of the order of 10 nm would be achieved between the filaments. If the filament ends were fixed, then tension would develop from the resultant strain in the crossbridge: to what extent the crossbridge would move under such isometric conditions depends on the location of the compliance in the system. Direct evidence for this swinging crossbridge model has proven exceedingly difficult to acquire. One reason for this may be the fact that only a fraction of the heads are attached at any one instant and during rapid shortening this may be a very small fraction indeed. Furthermore, the heads act asynchronously. Thus any structural signal is present on a high background from crossbridges in other intermediate states.

Myosin ATPase activity and actin activation

Myosin from vertebrate skeletal muscle hydrolyses MgATP slowly (k_{cat}=0.05 s^{-1}), but this activity is accelerated by two orders of magnitude by F-actin at low ionic strength. While such activation shows a roughly hyperbolic dependence on the actin concentration, interpretation of V_{max} and K_m is difficult because the actin and myosin sites are positioned on filamentous arrays that form a poorly organized meshwork, and hence the local concentrations and the availability of sites are unclear. For this reason most mechanistic studies have been performed using soluble myosin fragments (S1 and HMM). The simplest scheme that can describe the ATPase mechanism is given in eqn. (1) (where M is a myosin head):

$$M+ATP \rightleftharpoons M \cdot ATP+H_2O \rightleftharpoons M \cdot ADP \cdot P_i \rightleftharpoons M+ADP+P_i \qquad (1)$$

Although the steady-state ATPase activity of myosin is slow, the hydrolysis step itself is rapid (about 100 s^{-1}) and leads to a long-lived products complex (M·ADP·P$_i$) as the major steady-state intermediate. This was established using quenched-flow techniques to acid-denature the myosin after mixing with excess ATP on a millisecond time scale. A burst of P$_i$ and ADP production was observed, corresponding to the release of products from denatured M·ADP·P$_i$ formed during the first turnover, whereas quenching at longer times revealed additional products produced at the slow steady-state rate. The hydrolysis step is readily reversible and in rapid equilibrium so that a significant concentration (typically 10–20%) of the nucleotide is present as the M·ATP complex. Product release is limited by an isomerization step after which P_i is rapidly released and ADP somewhat slower. Intrinsic fluorescence and extrinsic probes demonstrate that the conformations of the M, M·ADP and M·ADP·P_i states differ. The M·ATP state is more difficult to characterize but may share some similarities with the M·ADP state.

During the steady-state turnover of ATP by a myosin head, the predominant intermediate is M·ADP·P_i but with significant contributions from M·ATP and M·ADP binary complexes, particularly at low temperatures. This mixture complicates the definition of structural studies and also, after several minutes, the ATP becomes exhausted, limiting the time for data capture. To overcome this problem analogue states have been developed. Simply adding high concentrations of P_i to M·ADP does not generate significant amounts of M·ADP·P_i; indeed, P_i tends to displace the ADP and forms a M·P_i complex. However, it turns out that a number of compounds with a similar size and shape to P_i will form a M·ADP·P_i-like state, notably VO_4^{3-}, AlF_4^- and BeF_3^-. A characteristic of these complexes is that they form slowly (a time scale in the order of minutes) but dissociate over several hours or days, giving ample time for structural measurements.

In the absence of nucleotide, S1 binds tightly to F-actin (K_d=10–100 nM) to give decorated filaments in which the S1 heads project at the characteristic rigor angle. Addition of ATP causes dissociation of the S1 heads but, somewhat paradoxically, the S1 ATPase activity is enhanced. The solution to this paradox came with Lymn & Taylor's [5] investigations comparing the acto-HMM dissociation rate constant, measured by using turbidity in a stopped-flow apparatus, with the hydrolysis rate constant measured by the quenched-flow technique. They concluded that the A·M·ATP complex (where A is actin) rapidly dissociates and that hydrolysis occurs predominantly on the dissociated myosin (i.e. the same M·ATP to M·ADP·P_i transition as in eqn. 1). In order to explain the activation of the ATPase, it was proposed that actin rebound to the M·ADP·P_i state to bypass the slow myosin pathway (eqn. 2).

$$A{\cdot}M+ATP \rightleftharpoons A{\cdot}M{\cdot}ATP+H_2O \rightleftharpoons A{\cdot}M{\cdot}ADP{\cdot}P_i \rightleftharpoons A{\cdot}M+ADP+P_i \qquad (2)$$

$$M{\cdot}ATP+H_2O \rightleftharpoons M{\cdot}ADP{\cdot}P_i$$

The dissociation and reassociation of the myosin head with actin during ATPase activity suggested a possible crossbridge cycle (Figure 2).

The Lymn–Taylor scheme remains the framework upon which many new results are discussed. However, it is an incomplete description of events. In

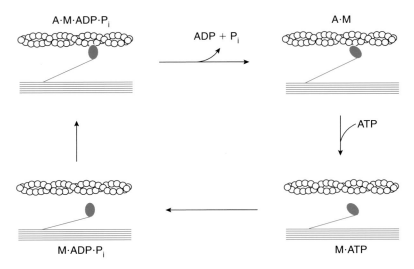

Figure 2. The Lymn–Taylor scheme for mechanochemical coupling [5]
ATP dissociates the rigor complex (A·M). Hydrolysis of bound ATP occurs primarily on the detached head (M·ATP↔M·ADP·P$_i$), whereas ADP and P$_i$ release is induced by actin rebinding. The scheme proposed originally suggested that the crossbridge stroke arises from tilting of the attached head, whereas recent data suggest it may bend to effect a similar translation of about 10 nm for each cycle.

their original work, Lymn & Taylor [5] concluded that at high actin concentrations release of P$_i$ from the A·M·ADP·P$_i$ complex was the major rate-limiting step. Subsequently it was shown under some conditions that hydrolysis may become rate limiting, although in myofibrils held under isometric conditions, the overall turnover rate is slower than with isolated proteins and is limited by P$_i$ release [6]. This link between structure and chemistry is crucial and follows from thermodynamics: if the free energy of ATP hydrolysis is converted to mechanical work, then imposing a load on a muscle should affect the equilibrium, and hence kinetics, of the ATPase reaction.

These schemes (eqns. 1 and 2) have been expanded to incorporate changes in myosin conformation revealed by transient-state kinetics. In an attempt to simplify the relation between structure and nucleotide state, Shriver [7] analysed models in which there were two fundamental conformations of the myosin, and the equilibrium between the two was shifted by nucleotide binding. This concept was supported by spectroscopic evidence and seems to be in accord with recent X-ray crystallographic results (see below). However, Shriver [7] concluded that the degree to which the conformational state could be switched by nucleotide binding was insufficient to account for the observed changes in actin affinities in a thermodynamically self-consistent way, i.e. there are further conformational transitions that modulate actin affinity.

One of the crucial tests of the Lymn–Taylor model (Figure 2) is to gain structural information about the A·M·ADP·P$_i$ state prior to the putative crossbridge swing. Unfortunately analogues that favour such a state tend to result in

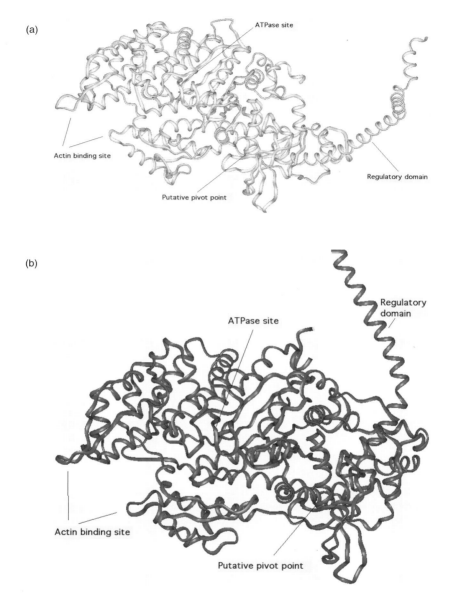

Figure 3. Comparison of the structures of skeletal-muscle S1 in its nucleotide-free form (a) with smooth-muscle S1 in the ADP·AlF$_4$-bound state (b), with the conserved β-sheet at the active site co-aligned to reveal about a 70° tilt of the regulatory domain

The light chains have been left out for clarity. Data are from Brookhaven files 2MYS and 1BR1 and [3,10].

actin dissociation. Working at very low ionic strength helps to counteract this, but significant dissociation may still occur. On the other hand, ADP is less efficient at dissociating actin and, under appropriate conditions, a significant concentration of the ternary A·M·ADP state can be generated, particularly with smooth-muscle proteins.

X-ray crystallographic studies

Crystallographic studies of actin and myosin present a challenge because both proteins spontaneously polymerize. In the case of actin, the crystal structure of the G form was solved in combination with other proteins that keep it in the monomeric form (e.g. DNase, profilin, gelsolin). The structure of F-actin was then modelled by fitting the high resolution structure of G-actin into the electron density envelope observed by electron microscopy [8].

Fragments of myosin, generated by proteolysis or recombinant DNA technology, have proven amenable to crystallization. To date, the key structures solved are chicken skeletal S1 (Figure 3a), containing both light chains and a sulphate ion (but no nucleotide) at the active site [3], a number of truncated *Dictyostelium* S1 constructs containing various nucleotides and analogues at the active site but lacking light chains [9], and a recombinant chicken smooth-muscle S1 containing an essential light chain and bound nucleotide [10]. Overall, the folds of the motor domain are very similar, and indeed the active site shares conserved motifs with kinesins and G-proteins [11]. In particular, the nucleotide-binding site contains three loops (the P-loop, switch 1 and switch 2), which form specific interactions with the triphosphate moiety.

Interestingly, structures of the *Dictyostelium* constructs with a variety of ligands fall into two classes that differ in the position of the switch-2 loop. Nucleotide-free states and analogues thought to mimic the M·ATP state (M·ADP and M·ATPγS) resemble the original skeletal S1 structure. However, with analogues thought to mimic a transition state approaching the M·ADP·P$_i$ intermediate (ADP plus AlF$_4^-$ or VO$_4^{3-}$), the switch-2 loop moves closer to the putative position of the γ-phosphate. This movement is coupled to the closure of the cleft between the upper and lower 50 kDa domains, and a rearrangement of the C-terminus. The latter suggested that the regulatory domain, missing in these structures, would project at a very different angle, but such speculation was tempered by the possibility that truncation induced an artefactual rearrangement of the C-terminus.

Fortunately this situation was clarified with the solution of the smooth-muscle S1·ADP·AlF$_4^-$ structure (Figure 3b), where the regulatory domain was indeed rotated about 70° relative to its position in the nucleotide-free skeletal S1 [10]. This movement, which pivots about a point in the 20 kDa domain, corresponds to about a 10 nm displacement at the distal end of the regulatory domain, where it connects to S2 in the intact myosin molecule. Thus the regulatory domain, aside from acting as a control element in some myosins, may function as a lever arm. However, this study also revealed a complication in that the M·ADP·BeF$_3$ structure with smooth-muscle S1 also took on the bent structure, whereas with the *Dictyostelium* construct, this analogue resembled the extended skeletal myosin structure. Perhaps this result is not unexpected in view of earlier solution studies. The M·ATP and M·ADP·P$_i$ states differ very little in energy so that crystal forces could easily select one form over the other. Furthermore, reference was made above to the fact that any single nucleotide state may exist in two or more conformations in solution [7], which would be free to re-equilibrate as crystallization proceeded.

The location of the actin–myosin binding interface has been defined by fitting the crystal structures of G-actin and the myosin head into the envelope of density defined by electron microscopy of S1-decorated actin filaments. The overall fit suggests the head must be in the extended conformation close to that of skeletal S1 in the absence of bound nucleotide. Whereas the residues that are located at the interface can be identified from the fitting procedure, the precise contacts cannot be identified. The actomyosin interface is clearly multivalent, with electrostatic interactions predominately involved in a weak, non-stereo specific interaction, whereas hydrophobic interactions may drive the rearrangement to a more stable attachment. However, this interface must change in a dramatic way (in energetic terms) when nucleotide is bound to the active site, even if the structural reorganization is rather small. Communication between the nucleotide and actin sites may occur via the opening and closing of a cleft that runs between the so-called 50 kDa domain. Movement of the switch-2 loop in the M·ADP·AlF$_4$ structure is coupled to the closing of the cleft, and could account for its weak actin-binding properties. However, the more open state of the cleft observed with other nucleotides (e.g. ATPγS) is comparable in gross structure with that of the nucleotide-free state, yet the binding affinities for actin differ by several orders of magnitude.

Overall, crystallographic studies support the idea that the myosin head can bend, and if such a movement is reversed when it is attached to actin, it would account for a stroke size of about 10 nm in the correct direction. Furthermore, only about 35% of the myosin mass moves in this transition, which may account for the difficulty of resolving this process by fibre diffraction from whole muscle. However, our understanding of the effects of actin is incomplete.

Evidence for lever-arm movement in solution studies

The characterization of the lever-arm positions in the crystal structures begs the questions: 'does such a change occur in solution?' or, conversely, 'is the change an artifact of crystallization?' and 'does it occur when the myosin head is attached to actin?'. In fact, the question of whether the myosin head bends was asked long before the recent crystallographic results.

Solution measurements on isolated S1 heads using electric birefringence [12] and X-ray scattering [13] indicated a shift in conformation to a more compact form in the presence of bound nucleotide. Electron microscopy has also revealed that S1 heads can take on straight or bent conformations, although the nature of the coupling to the nucleotide state is controversial (compare [14] and [15]). Recently, a fusion protein has been made in which green fluorescent protein (GFP) was attached to the N-terminus and blue fluorescent protein (BFP) was attached to the C-terminus of the motor domain in place of the regulatory domain (Figure 4). This construct displays fluorescence resonance energy transfer (FRET) in which excitation of BFP results in dipolar coupling with GFP and emission of fluorescence characteristic of the latter, a phenomenon that is distance- and orientation-dependent. During ATP turnover the FRET efficiency between BFP and GFP changes, indicating a movement or reorientation of the C-terminal region relative to the N-terminus [16]. Interestingly, BeF_3^- (which may promote either of the two crystal conformations) gave FRET efficiencies halfway between those of the M and $M \cdot ADP \cdot P_i$ states. The GFP approach opens the way to examining such conformational changes in the presence of actin.

Cross-linking studies between thiol residues on a helix at one end of the putative pivot point (see Figure 3) indicate a variable distance between the cysteine residues dependent on the state of the bound nucleotide [17,18]. The crystal structures described above both have an intact helix in which the thiols are about 1.8 nm apart. However, most recently, a third conformation has been solved by crystallography in which the helix is melted and the regulatory domain straightened compared with the original chicken structure [19].

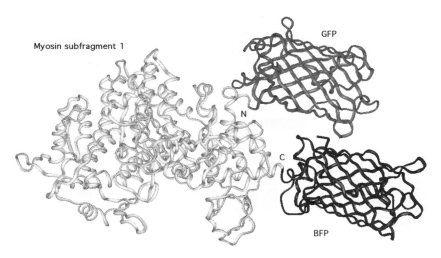

Myosin subfragment 1

GFP

N

C

BFP

Figure 4. The use of GFP–BFP FRET pairs to follow conformational changes in the neck region of the myosin molecule [16]
ATP binding induces an isomerization that leads to a reorientation between the N- and C-termini and relative movement of the GFP and BFP fluorophores (located within their β-cage structure). The transition is reversed on P_i release [16]. The diagram was constructed using co-ordinates from Brookhaven files 2MYS and 1EMB.

Evidence that the regulatory domain acts as a lever has come from motility assays *in vitro* using constructs that have artificially lengthened or shortened necks. In general there is a good correlation between the length of the lever arm and the observed sliding velocity [20]. Myosins in different classes show a wide range of lever-arm lengths, as judged by the number of light-chain-binding sites (one to six) and these may be adaptations to provide variable gearing to the task at hand.

Examination of head movements while attached to actin is more difficult because only a small fraction of the heads may be involved. In solution, high actin concentrations are required to favour binding, so in many cases it is more convenient to use permeabilized muscle fibres, which come with the added advantage of a high degree of alignment. As a test of the Lymn–Taylor model, many early studies attempted to monitor angular changes in the myosin head by attaching optical or magnetic probes to the motor domain, via such groups as the reactive thiols. In general, little evidence was produced for head attachment at a unique angle which differed from that of rigor, although it was considered that heads may attach over a range of angles, based on estimates of the fraction of heads attached. More recently, probes have been introduced via the light chains to test for a swing of the lever arm [21]. Small fluorescence polarization signals have been recorded from a rhodamine probe on the regulatory light chain that may be explained by a few heads (10–15%) undergoing a large angle change (>30°). However, correlation with chemical states is difficult because different heads in the myofibril array experience different mechanical states.

Problems and prospects

The fundamental concepts of the swinging crossbridge model (i.e. physical attachment of the myosin head to actin followed by some form of structural rearrangement) have been with us for 40 years. These basic tenets have been challenged by alternative hypotheses, including ones that place the contractile event in myosin S2 or, more radically, have no direct physical link between actin and myosin and rely on indirect effects of long-range charges or water movements within the myofibril lattice. The recent demonstration of force generation by a single myosin head lacking S2 and an organized myosin lattice, and the direct observation of the coupling between head attachment to actin and a single ATP-turnover cycle, lend powerful support to the classical crossbridge theory [22]. These ideas are also supported by the discovery of the myosin superfamily of proteins, which retain a conserved motor domain, but are otherwise very diverse, and in some cases function as individual porters rather than in a filamentous array. Crystallography is now beginning to define the likely movements that the myosin head can make. The communication path between events at the active site and both the proximal actin-binding site and the distal regulatory domain are now being defined at the atomic level.

Site-directed mutagenesis studies are providing critical tests for these ideas. However, these studies are still incomplete. In particular, events at the actin interface and conformational changes within actin itself are unclear.

Whereas there is now direct evidence that the turnover of a single ATP molecule by a singe S1 head can generate a translational movement (5–15 nm) comparable with that predicted by the swing or bend of the head itself, there remain some unexplained discrepancies. In particular, under conditions of low load, the overall ATPase rate and the working stroke size argue for a very low duty ratio if the mechanics and ATPase are coupled in a 1:1 fashion [1,2]. During sliding at maximum velocities, only about 1% of the heads can undergo this stroke at a particular instant. It is possible that some heads may attach and use their binding energy to impart an impulse to the actin, only to be ripped off by other heads before completing their ATPase cycle. Thus the number of attached heads may be higher than 1%. By 'living on borrowed energy' the free energy of ATP hydrolysis may be partitioned over several heads [2]. Evidence for multiple strokes has been obtained within muscle fibres by imposing a rapid staircase of releases. Nevertheless, such a mechanism does not account for the results from some motility assays *in vitro* where the total number of available actin and myosin sites are limited [23].

Summary

- *Solution measurements indicate that actin and myosin alternately bind and dissociate during one ATP hydrolysis cycle.*
- *Crystallographic studies indicate at least two basic conformations of the myosin head exist in which the regulatory domain swings through an angle of about 70°.*
- *Actin must further modulate these conformations, but high-resolution information about the actomyosin interface is lacking.*
- *One-to-one coupling between the ATPase cycle and a mechanical cycle involving myosin-head bending could account for about a 10 nm stroke size.*
- *At high sliding velocities, discrepancies remain which suggest that a myosin head may undergo repetitive interactions with actin for each ATP hydrolysed.*

I am grateful to Dr. M.J. Sutcliffe for assistance with molecular graphics and Dr. P.B. Conibear for discussions. Financial support was provided by The Royal Society and the Wellcome Trust.

References

1.   Bagshaw, C.R. (1993) *Muscle Contraction*, 2nd edn, Chapman and Hall, London
2.   Cooke, R. (1997) Actomyosin interaction in striated muscle. *Physiol. Rev.* **77**, 671–697

3. Rayment, I., Rypniewski, W.R., Schmidt-Bäse, K., Smith, R., Tomchick, D.R., Benning, M.M., Winkelmann, D.A., Wesenberg, G. & Holden, H.M. (1993) Three-dimensional structure of myosin subfragment-1: a molecular motor. *Science* **261**, 50–58

4. Hirose, K., Lenart, T.D., Murray, J.M., Franzini-Armstrong, C. & Goldman, Y.E. (1993) Flash and smash: rapid freezing of muscle fibers activated by photolysis of caged ATP. *Biophys. J.* **65**, 397–408

5. Lymn, R.W. & Taylor, E.W. (1971) Mechanism of adenosine triphosphate hydrolysis by acto-myosin. *Biochemistry* **10**, 4617–4624

6. Lionne, C., Brune, M., Webb, M.R., Travers, F. & Barman, T. (1995) Time resolved measurements show that phosphate release is the rate limiting step on myofibrillar ATPases. *FEBS Lett.* **364**, 59–62

7. Shriver, J.W. (1986) The structure of myosin and its role in energy transduction in muscle. *Biochem. Cell. Biol.* **64**, 265–276

8. Holmes, K.C., Popp, D., Gebhard, W. & Kabsch, W. (1990) Atomic model of the actin filament. *Nature (London)* **347**, 37–44

9. Gulick, A.M. & Rayment, I. (1997) Structural studies on myosin II: communication between distant protein domains. *BioEssays* **19**, 561–569

10. Dominguez, R., Freyzon, Y., Trybus, K.M. & Cohen, C. (1998) Crystal structure of a vertebrate smooth muscle myosin motor domain and its complex with the essential light chain: visualization of the pre-power stroke state. *Cell* **94**, 559–571

11. Kull, F.J., Vale, R.D. & Fletterick, R.J. (1998) The case for a common ancestor: kinesin and myosin motor proeins and G-proteins. *J. Muscle Res. Cell Motility* **19**, 877–886

12. Highsmith, S. & Eden, D. (1993) Myosin-ATP chemomechanics. *Biochemistry* **32**, 2455–2458

13. Wakabayashi, K., Tokunaga, M., Kohno, I., Sugimoto, Y., Hamanaka, T., Takezawa, Y., Wakabayashi, T. & Amemiya, Y. (1992) Small-angle synchrotron X-ray scattering reveals distinct shape changes of the myosin head during hydrolysis of ATP. *Science* **258**, 443–447

14. Burgess, S.A., Walker. M.L., White, H.D. & Trinick, J. (1997) Flexibility within myosin heads revealed by negative stain and single-particle analysis. *J. Cell. Biol.* **139**, 675–681

15. Bagshaw C.R. (1997) Myosin trapped but not tamed. *Nature (London)* **390**, 345–346

16. Suzuki, Y., Yasunaga, T., Ohkura, R., Wakabayashi, T. & Sutoh, K. (1998) Swing of the lever arm of the myosin motor at the isomerization and phosphate-release steps. *Nature (London)* **396**, 380–383

17. Nitao, L.K. & Reisler, E. (1998) Probing the conformational states of the SH1-SH2 helix in myosin: a cross-linking approach. *Biochemistry* **37**, 16704–16710

18. Liang, W. & Spudich, J.A. (1998) Nucleotide-dependent conformational change near the fulcrum region in *Dictyostelium* myosin II. *Proc. Natl. Acad. Sci. U.S.A.* **95**, 12844–12847

19. Houdusse, A., Kalabokis, V.N., Himmel, D., Szent-Györgyi, A.G. & Cohen, C. (1999) Atomic structure of scallop myosin subfragment S1 complexed with MgADP: a novel conformation of the myosin head. *Cell* **97**, 459–470

20. Uyeda, T.Q.P., Abrahamson, P.D. & Spudich, J.A. (1996) The neck region of the mysoin motor domain acts as a lever arm to generate movement. *Proc. Natl. Acad. Sci. U.S.A.* **93**, 4459–4464

21. Goldman, Y.E. (1998) Wag the tail: structural dynamics of actomyosin. *Cell* **93**, 1–4

22. Ishijima, A., Kojima, H., Funatsu, T., Tokunaga, M., Higuchi, H., Tanaka, H. & Yanagida, T. (1998) Simulatneous observation of individual ATPase and mechanical events by a single myosin molecule during interaction with actin. *Cell* **92**, 161–171

23. Saito. K., Aoki, T., Aoki, T. & Yanagida, T. (1994) Movement of single myosin filaments and myosin step size on an actin filament suspended in solution by a laser trap. *Biophys. J.* **66**, 769–777

4

Unconventional myosins

Georg Kalhammer* & Martin Bähler†[1]

*Adolf-Butenandt-Institut, Zellbiologie, Ludwig-Maximilians-Universität, Schillerstr. 42, D-80336 München, Germany and †Westf. Wilhelms-Universität Münster, Institut für Allgemeine Zoologie und Genetik, Schlossplatz 5, D-48149 Münster, Germany

Introduction

Living organisms exhibit a wide range of motility phenomena, with muscle contraction being one of the most obvious. But in addition to muscle contraction, there exist many other cellular and intracellular forms of motility that include cell migration, cell-shape changes and intracellular transport. Situations in which these forms of motility are involved in higher organisms include wound healing, immune defence, cytokinesis, morphological differentiation of cells, and transport of organelles and various subcellular components. These activities are coupled to a dynamic cytoskeleton consisting of different filamentous structures such as actin filaments, microtubules and intermediate filaments. It is these filaments that serve as tracks for molecular motors.

Recently, many putative molecular motors that use actin filaments as tracks for unidirectional force production have been identified. These molecular motors share a conserved motor-domain sequence that defines them as members of the myosin superfamily. They are hetero-oligomers consisting of one or two heavy chains and a variable number of light chains (Figure 1). The heavy-chain amino acid sequence can be divided into different regions. The motor region is formed by a globular head domain that binds actin and hydrolyses ATP in an F-actin-regulated manner, and a neck domain that con-

[1]To whom correspondence should be addressed.

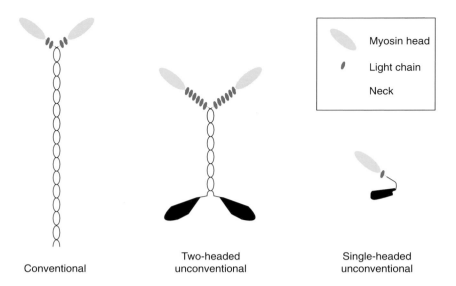

Conventional   Two-headed   Single-headed
                             unconventional   unconventional

Figure 1. Schematic representations of myosin structures
Myosin molecules consist of a head domain, a neck domain of variable length with different numbers of light chains associated, and a tail domain. The tail domains of conventional (class-II) and
some unconventional myosins dimerize and form an α-helical coiled-coil structure, giving rise to
double-headed molecules. However, the tail domains of many unconventional myosins do not
dimerize and these myosins are therefore single headed.

tains a variable number of light-chain-binding motifs. The number of light-
chain-binding motifs in the neck domain can vary from one to seven. These
motifs are also termed 'IQ motifs' based on the presence of the conserved
amino acids isoleucine and glutamine. Light chains belong to the
calmodulin/EF-hand superfamily of calcium-binding proteins. They are
thought to provide rigidity to the neck region (that acts as a lever arm) and to
regulate the motor activity (often via the messenger Ca^{2+}). Different lever
lengths may allow the different myosins to advance by different step sizes
along actin filaments. Finally, the myosins contain a tail region that can be very
divergent. Some myosins contain in their tail regions sequences that allow
dimerization through the formation of coiled-coil structures. Conventional
myosin (Figure 1), the best-studied form of myosin, dimerizes via long tail
regions which then assemble further to form filaments. Well-known examples
of conventional myosins (now called class-II myosins) are muscle myosins (see
Chapter 3 in this volume) and cytoplasmic myosins from unicellular organisms. All myosins that cannot form filaments are called 'unconventional'
myosins.

The myosin superfamily

Based on comparison of their head-domain sequences, myosins are grouped
into different classes. During the last decade many novel myosin molecules

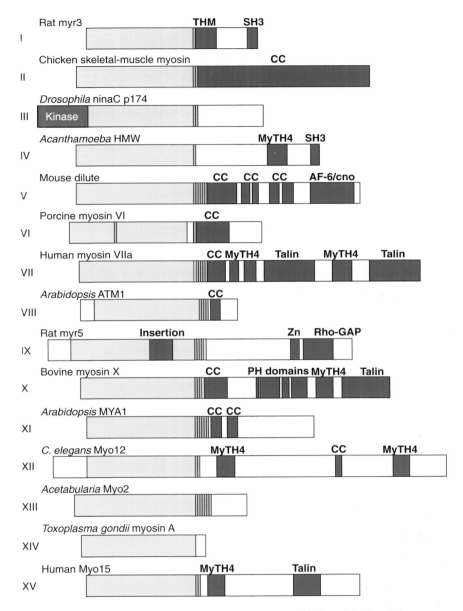

Figure 2. Schematic representations of the diverse motifs found in the different myosins

Examples are shown for each of the myosin classes (I–XV) from a variety of different organisms. The myosin head domains are coloured light blue and the light-chain-binding motifs are coloured grey. In dark blue are diverse sequence motifs, as indicated. THM, myosin-I tail homology motif; SH3, Src homology 3 domain; CC, α-helical coiled-coil domain; Kinase, Ser/Thr protein-kinase domain; AF-6/cno, AF-6/canoe homology domain; MyTH4, myosin tail homology 4 domain; Talin, talin homology domain; Insertion, large insertion within the myosin head domain at an actin-binding surface loop; Zn, Cys_6His_2 zinc-ion-binding domain; Rho-GAP, Rho GTPase-activating protein domain; PH domains, pleckstrin-homology domains.

have been identified, and 15 different classes (I–XV) are known so far. In the tail regions of myosin molecules one can find motifs and domains known from other proteins (Figure 2). But even the head domains of some myosins contain insertions or/and N-terminal extensions.

It has been shown that many myosins from different classes and even several isoforms of each class can be expressed in a single cell. This finding raises the question of what are the cellular functions of this multitude of myosins. Furthermore, to function in a useful manner, these myosins are likely to be regulated differently. Comparisons of the different properties of such a great number of different myosin molecules also offers the possibility of obtaining more information about the molecular mechanisms of mechanochemical energy transduction. In the following sections, we will describe some of the known basic functions of several myosin classes. Some myosins may have more than one function.

Class-V myosins are involved in the transport of organelles and cellular particles along actin filaments

It was suggested some time ago that myosins move vesicles and organelles along actin filaments. Vesicles and organelles in axoplasm extruded from the squid giant axon have been observed moving along actin filaments [1]. However, transport along actin filaments by a myosin requires that it can move a certain distance along the actin filament without detaching and diffusing away. This is not the case for conventional myosin, which spends only a short fraction ($\approx$5%) of its mechanochemical cycle strongly attached to actin filaments. Therefore, to get continous movement along the actin filament it is likely that several myosins need to co-operate with each other, so that at any particular time at least one myosin is holding on to the actin filament. It remains to be seen whether single unconventional myosins can move continously along actin filaments or whether they work in assemblies of several molecules. However, it seems likely that many unconventional myosins will not serve transport functions.

For some time, class-V myosins have been the prime candidates for vesicle and organelle motility. Myosins of this class have in their tail domains one or more regions that are predicted to form an α-helical coiled-coil structure. Indeed, it has been shown by electron microscopy that purified brain myosin Va exists as a two-headed dimer [2]. Six light chains are bound to its neck domain. Three class-V myosins from mammals, two from baker's yeast and one each from chicken, *Caenorhabditis elegans* and *Drosophila melanogaster* have been identified so far.

Evidence that myosin V serves a transport function is emerging from biochemical studies and from studies on mutant mice and yeast. Myosin Va purified from chick brain can translocate beads along actin filaments *in vitro* [3] and activated brain myosin V associated with vesicles will support vesicle transport along actin filaments [4,5]. It binds more tightly to actin filaments in the presence of ATP than conventional myosin [6]. Therefore, it could be bound to the actin filament for a larger fraction of the cycle time. Whether it really acts as a processive motor like kinesin (see Chapter 6 in this volume) still remains to be shown conclusively.

Mice defective in the myosin Va (or *dilute*) gene exhibit a washed out or 'diluted' coat colour. In addition, affected mice exhibit severe neurological problems and die a few weeks after birth. The diluted coat colour is due to a defective transfer of melanin pigment from the melanocytes to the hair shaft. Myosin Va has been shown to be associated with pigment-containing melanosomes. The melanosomes are clustered in the cell body of melanocytes in *dilute* mice whereas they are distributed in the dendritic processes of wild-type melanocytes [7]. This finding is in agreement with a role for myosin Va in melanosome transport to the cell periphery or in tethering of melanosomes in the periphery.

The budding yeast *Saccharomyces cerevisiae* possesses two genes encoding class-V myosin heavy chains called *MYO2* and *MYO4*. Myo2p is essential for polarized growth of yeast cells. At the restrictive temperature, cells carrying a temperature-sensitive allele of *MYO2* form abnormally large mother cells and very small buds, demonstrating a defect in the localization of growth. After completion of budding, cells are arrested as large unbudded cells. Incubation at the restrictive temperature also leads to the accumulation of small vesicles in the cells [8,9]. These results are suggestive of a role for Myo2p in vectorial transport of vesicles to the bud. In accordance with this notion, vacuole inheritance is also impaired at the restrictive temperature [10]. Yeast Myo4p is required for correct localization of ASH1 (asymmetric synthesis of homothallism 1) mRNA particles to the bud tip [11]. Ash1p in turn suppresses in the daughter cell the expression of homothallism endonuclease that is important for mating-type switching in yeast [12]. Therefore, the transport of the ASH1 mRNA is a prerequisite for asymmetric cell division. Myo4p represents the prime candidate for transporting mRNA particles to the tip of the bud.

Taken together, these many lines of evidence argue for class-V myosins being involved in moving organelles and particles along actin filaments.

Class-I myosins are implicated in membrane dynamics

Regions of the plasma membrane of cells can be arranged into functionally highly specialized structures that are either dynamic or relatively stable. For example, microvilli on certain epithelia have to keep their organization to ensure proper function of the epithelium. In contrast, phagocytosis and pinocytosis involve the rapid formation of either membrane protrusions or invaginations to engulf particles and take up solutes. Several myosins have been implicated in either the maintenance or rearrangement of such membranous specializations.

Class-I myosins were described first in the soil amoeba *Acanthamoeba castellanii*. These amoeboid myosins I were found to have a motor domain, a single light-chain-binding site (IQ motif), and a relatively short basic tail domain that encompasses an SH3 (Src homology 3) domain. SH3 domains are thought to represent protein–protein-interaction motifs and are found in many proteins, especially in proteins involved in signal transduction and in proteins associated with the actin cytoskeleton. More recently, homologues of class-I myosins have been identified in many organisms, including vertebrates. Some of these lack an SH3 domain and some have several light-chain-binding sites. But all class-I myosins are monomeric, single-headed molecules [13].

The class-I myosin 'brush-border myosin I' is expressed specifically in intestinal brush-border cells where it links the central actin-filament core of microvilli to the plasma membrane [13]. There it appears to play a role in maintaining the microvillar structure. Similarly, ninaC p174 (ninaC standing for neither inactivation nor afterpotential C), a class-III myosin, is required for the maintenance of microvillar rhabdomeres in the *Drosophila* eye. Deletion of or mutations in ninaC p174 cause the degeneration of the rhabdomeres [14].

The budding yeast *S. cerevisiae* possesses two genes encoding myosin-I heavy chains, *MYO3* and *MYO5*. Deletion of *MYO3* does not yield an obvious phenotype. Cells lacking *MYO5* exhibit a defect in the internalization step of endocytosis (as measured by α-factor internalization), but grow at a normal rate [15]. However, *MYO3/MYO5* double-mutant cells have a strong growth defect, loss of actin polarity and polarized cell-surface growth, and an accumulation of intracellular membranes [16]. This points to some redundancy in the functions of these two myosins that might be a general problem in the analysis of myosin functions. This is also supported by results obtained in the slime mould *Dictyostelium discoideum*, where six class-I myosins have been identified. Knock-out of single myosin-I genes generally results in no (or only subtle) defects in cell motility, pinocytosis and phagocytosis. Upon removal of a particular myosin I, the cells exhibit more pseudopods than wild-type cells, which leads to more frequent turning and lower net movement. In double or triple mutants, pinocytosis is decreased further; since mutant cells form even more actin-rich pinocytic crowns than wild-type cells, this defect is likely to be due to an inability to retract these membrane protrusions [17].

A class-I myosin from the fungus *Aspergillus nidulans*, MyoA, is necessary for the establishment of polarized growth of the fungal hyphae. Interestingly, replacement of the endogenous *MYOA* gene with a gene coding for a constitutively activated version of MyoA resulted in the accumulation of membranes in the growing hyphae. This phenomenon is explained by an activation of endocytosis [18].

Together these data argue strongly for a role of class-I myosins and possibly myosin III in membrane dynamics associated with organization of the cell cortex.

Directed signal transduction

The activity of myosin molecules is assumed to be controlled and regulated in cells. However, myosin molecules are not only regulated themselves; recent data demonstrate that they are able to regulate the activity of other molecules that serve a role in signal transduction. Some myosins (classes III and IX) even exhibit activities characteristic of signalling molecules. For instance, the class-III myosins contain, N-terminal to the myosin head domain, a Ser/Thr protein-kinase domain (Figure 2) that can phosphorylate several substrates *in vitro* [19]. Although the relevant substrate(s) *in vivo* have not yet been identified, it is clear that a functional kinase domain is necessary for the proper function *in vivo* of the class-III myosin ninaC p174 from *D. melanogaster*. This myosin is expressed exclusively in the photoreceptor cells of the eye and is localized in the rhabdomeres, the sites of light-information processing. Its kinase domain is required for normal electrophysiological responses to light stimuli [14].

Many cellular signalling systems involve small G-proteins, such as the small Ras-related G-protein, Rho. These small G-proteins act as molecular switches, being active in their GTP-bound state and inactive in their GDP-bound state. In their active GTP-bound conformation, they transmit signals to their downstream targets. GTP can be hydrolysed by the G-protein itself (they are GTPases), but this intrinsic hydrolysis is rather slow. The hydrolysis can be greatly accelerated by GTPase-activating proteins (GAPs), which thereby turn off the small G-proteins. Class-IX myosins such as rat myr5 (Figure 2), rat myr7 and human myosin IXb contain in their tail domains a Rho-GAP region that can activate the GTPase of the small G-protein Rho [20]. Interestingly, the small G-protein Rho that is negatively regulated by class-IX myosins plays a major role in controlling the organization of the actin cytoskeleton [21]. Activation of Rho via extracellular signals induces the formation of actin stress fibres and actomyosin-driven cell contraction (Figure 3). The molecular mechanism involves the activation of a Rho kinase by active Rho, which in turn leads to the activation of conventional non-muscle myosin through the phosphorylation of its regulatory light chains. Note that in class-IX myosins the myosin domain depends for its function(s) on the organization of the actin filaments, whereas the tail domain negatively regulates the activity of the small G-protein Rho that controls the actin-filament organization. Future experiments will have to determine the exact molecular functions of class-IX myosins in the Rho signalling pathway.

Two other Rho family members, Cdc42 and Rac, might also regulate myosin function (Figure 3). Amoeboid class-I myosins are activated by phosphorylation of their myosin head domains at distinct serine or threonine residues [22]. The myosin head-domain kinases responsible for this have been identified in *Acanthamoeba* and *Dictyostelium*. They show sequence homology to the PAK (p21 Cdc42/Rac-activated protein kinase)/Ste20 kinase family.

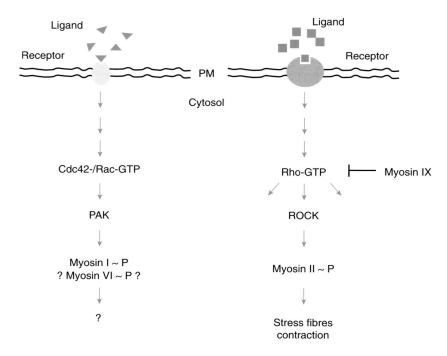

Figure 3. Myosin molecules in signal-transduction pathways of Rho-family small G-proteins

Small G-proteins of the Rho family are activated by extracellular signals (Ligand) that bind to receptors. These receptors initiate an intracellular signal cascade that activates the exchange of GDP for GTP on the Rho-family small G-proteins. The active GTP-bound Cdc42 or Rac binds to the PAK (p21 Cdc42/Rac-activated protein kinase) family of protein kinases that become activated and phosphorylate amoeboid myosin I and possibly myosin VI. The active GTP-bound Rho activates ROCK kinase (Rho-associated coiled-coil-forming protein kinase) that either phosphorylates the regulatory light chain of conventional myosin (class-II) directly and/or phosphorylates and inhibits the myosin phosphatase, leading indirectly to an increase in regulatory light-chain phosphorylation, thereby activating myosin II. Myosin IX serves as a negative regulator of Rho by stimulating the hydrolysis of GTP. PM, plasma membrane.

Indeed, PAK from rat brain and Ste20 from yeast are able to phosphorylate *Dictyostelium* myosin ID at the same site as the endogenous amoeboid kinases [23]. These kinases from mammals and yeast are activated by the Rho family members Rac and Cdc42. These small GTPases bind in their active GTP-bound conformation to PAK and increase its kinase activity. Activation of Rac and Cdc42 also leads to the rearrangement of the actin cytoskeleton, exemplified by the formation of lamellipodia and filopodia, respectively [21]. Class-VI myosins share the phosphorylation-site sequence of amoeboid class-I myosins and therefore also represent candidate targets of Cdc42/Rac-activated kinases [24]. These results demonstrate a close relationship between myosins and signal-transduction pathways of small G-proteins of the Rho family and more general actin organization. Actin organization has been implicated in regulating many cellular activities.

Perspectives

The number of different myosins known has grown rapidly over the last few years. Myosins are largely defined by a conserved myosin head-domain sequence. Although for many of these myosins the analysis of their functions is only just beginning, some general themes about their cellular functions are emerging. It has become clear that myosins play diverse roles in many cellular processes, including vesicle/particle transport, control of cell shape, cell migration, organization of the actin cytoskeleton, endocytosis, exocytosis and signal transduction. It will be a challenge in the future to determine the exact molecular and cellular functions of each individual myosin. To achieve this goal we will have to address several questions, such as: (i) what are their mechanochemical properties?; (ii) what are the partners with which they interact directly? and (iii) how are their activities regulated? Studies on the mechanochemical properties of the more divergent myosins will contribute to a better understanding of how myosin motors work and answer the question of whether they all exhibit true motor activity.

Summary

- *Myosins constitute a large superfamily of F-actin-based motor proteins found in many organisms from yeast to humans. A phylogenetic comparison of their head sequences has allowed them to be grouped into 15 different classes.*
- *Unconventional myosins can be monomeric or dimeric, but are thought not to form filaments, unlike conventional myosin.*
- *The double-headed class-V myosins are good candidates for transporting vesicles, organelles and (mRNA) particles along actin filaments.*
- *Class-I myosins are involved in membrane dynamics and actin organization at the cell cortex, thus affecting cell migration, endocytosis, pinocytosis and phagocytosis.*
- *A class-III myosin from Drosophila is required for phototransduction and maintenance of the rhabdomere.*
- *Class-IX myosins negatively regulate the small G-protein Rho, a signalling molecule that regulates the organization of the actin cytoskeleton.*
- *Protein kinases that are regulated by members of the Rho small G-protein family regulate the motor activities of different myosins.*

References

1. Kuznetsov, S.A., Langford, G.M. & Weiss, D.G. (1992) Actin-dependent organelle movement in squid axoplasm. *Nature (London)* **356**, 722–725
2. Cheney, R.E., O'Shea, M.K., Heuser, J.E., Coelho, M.V., Wolenski, J.S., Espreafico, E.M., Forscher, P., Larson, R.E. & Mooseker, M.S. (1993) Brain myosin-V is a two-headed unconventional myosin with motor activity. *Cell* **75**, 13–23

3. Wolenski, J.S., Cheney, R.E., Mooseker, M.S. & Forscher, P. (1995) *In vitro* motility of immunoadsorbed brain myosin-V using a Limulus acrosomal process and optical tweezer-based assay. *J. Cell Sci.* **108**, 1489–1496

4. Rogers, S.L. & Gelfand, V.I. (1998) Myosin cooperates with microtubule motors during organelle transport in melanophores. *Curr. Biol.* **8**, 161–164

5. Evans, L.L., Lee, A.J., Bridgman, P.C. & Mooseker, M.S. (1998) Vesicle-associated brain myosin-V can be activated to catalyze actin-based transport. *J. Cell Sci.* **111**, 2055–2066

6. Nascimento, A.A.C., Cheney, R.E., Tauhata, S.B.F., Larson, R.E. & Mooseker, M.S. (1996) Enzymatic characterization and functional domain mapping of brain myosin-V. *J. Biol. Chem.* **271**, 17561–17596

7. Provance, Jr., D.W., Wei, M., Ipe, V. & Mercer, J.A. (1996) Cultured melanocytes from dilute mutant mice exhibit dendritic morphology and altered melanosome distribution. *Proc. Natl. Acad. Sci. U.S.A.* **93**, 14554–14558

8. Johnston, G.C., Prendergast, J.A. & Singer, R.A. (1991) The *Saccharomyces cerevisiae* MYO2 gene encodes an essential myosin for vectorial transport of vesicles. *J. Cell Biol.* **113**, 539–551

9. Govindan, B., Bowser, R. & Novick, P. (1995) The role of Myo2, a yeast class V myosin, in vesicular transport. *J. Cell Biol.* **128**, 1055–1068

10. Hill, K.L., Catlett, N.L. & Weisman, L.S. (1996) Actin and myosin function in directed vacuole movement during cell division in *Saccharomyces cerevisiae*. *J. Cell Biol.* **135**, 1535–1549

11. Bertrand, E., Chartrand, P., Schaefer, M., Shenoy, S.M., Singer, R.H. & Long, R.M. (1998) Localization of ASH1 mRNA particles in living yeast. *Mol. Cell* **2**, 437–445

12. Bobola, N., Jansen, R.-P., Shin, T.H. & Nasmyth, K. (1996) Asymmetric accumulation of Ash1p in postanaphase nuclei depends on a myosin and restricts yeast mating-type switching to mother cells. *Cell* **84**, 699–709

13. Mooseker, M.S. & Cheney, R.E. (1995) Unconventional myosins. *Annu. Rev. Cell Dev. Biol.* **11**, 633–675

14. Porter, J.A. & Montell, C. (1993) Distinct roles of the Drosophila ninaC kinase and myosin domains revealed by systematic mutagenesis. *J. Cell Biol.* **122**, 601–612

15. Geli, M.I. & Riezman, H. (1996) Role of type I myosins in receptor-mediated endocytosis in yeast. *Science* **272**, 533–535

16. Goodson, H.V., Anderson, B.L., Warrick, H.M., Pon, L.A. & Spudich, J.A. (1996) Synthetic lethality screen identifies a novel yeast myosin I gene (MYO5): myosin I proteins are required for polarization of the actin cytoskeleton. *J. Cell Biol.* **133**, 1277–1291

17. Ostap, E.M. & Pollard, D.S. (1996) Overlapping functions of myosin-I isoforms? *J. Cell Biol.* **133**, 221–224

18. Yamashita, R.A. & May, G.S. (1998) Constitutive activation of endocytosis by mutation of myoA, the myosin I gene of *Aspergillus nidulans*. *J. Biol. Chem.* **273**, 14644–14648

19. Ng, K.P., Kambara, T., Matsuura, M., Burke, M. & Ikebe, M. (1996) Identification of myosin III as a protein kinase. *Biochemistry* **35**, 9392–9399

20. Chieregatti, E., Gärtner, A., Stöffler, H.-E. & Bähler, M. (1998) Myr 7 is a novel myosin IX-RhoGAP expressed in rat brain. *J. Cell Sci.* **111**, 3597–3608

21. Hall, A. (1998) Rho GTPases and the actin cytoskeleton. *Science* **279**, 509–514

22. Brzeska, H. & Korn, E.D. (1996) Regulation of class I and class II myosins by heavy chain phosphorylation. *J. Biol. Chem.* **271**, 16983–16986

23. Wu, C., Lee, S.-F., Furmaniak-Kazmierczak, E., Cote, G.P., Thomas, D.Y. & Leberer, E. (1996) Activation of myosin-I by members of the Ste20p protein kinase family. *J. Biol. Chem.* **271**, 31787–31790

24. Buss, F., Kendrick-Jones, J., Lionne, C., Knight, A.E., Cote, G.P. & Luzio, J.P. (1998) The localization of myosin VI at the Golgi complex and leading edge of fibroblasts and its phosphorylation and recruitment into membrane ruffles of A431 cells after growth factor stimulation. *J. Cell Biol.* **143**, 1535–1545

5

Muscle, myosin and single molecules

Alex E. Knight[1] & Justin E. Molloy

Biology Department, University of York, P.O. Box 373, York YO10 5YW, U.K.

Introduction

Historically, muscle physiologists began working with whole muscles and then progressed to making mechanical and biochemical measurements from individual muscle fibres and myofibrils. The development of motility assays *in vitro* went a step further by using the individual protein filaments (myosin and actin) that make up the sarcomere. The logical conclusion to this reductionist approach is the development of single-molecule technologies, which have recently enabled measurement of the individual protein–protein interactions that underlie muscle contraction. Results from such experiments have provided insights into the mechanism of actomyosin and other molecular motors. However, some of the results obtained by different laboratories seem contradictory and others at variance with conventional ideas of actomyosin force production.

In this chapter we hope to provide an introduction for the newcomer, highlighting problems of interpretation of single-molecule experiments and current areas of controversy. The scope of the chapter is limited to the study of actomyosin. However, these powerful techniques are also used to study other molecular motors such as kinesin, dynein, RNA polymerase, the bacterial flagellar motor and F_1-ATPase [1].

[1]*To whom correspondence should be addressed.*

Table 1. Comparing muscle myosin II and kinesin

Note that these are the extreme cases: there is a range of intermediate types of motor among both myosins and kinesins. For an excellent review of these concepts, see [20]. These properties are interrelated, and are to some extent determined by the structures of the track (the distance between motor-binding sites, or path distance) and of the motor (the working or power stroke). A motor that transports an organelle, especially if it acts individually, must remain attached to its track for long distances; it is a 'porter' [21] or processive motor. A motor that operates as part of a large 'team' must detach rapidly after its stroke to allow other motors to continue the sliding motion; it acts as a 'rower' and is not processive.

Characteristic	Myosin II	Kinesin
Substrate	Actin filament	Microtubule
Function	Rapid sliding	Organelle transport
Duty ratio	Low (mostly detached)	High (mostly attached)
Processivity	Not processive — 'rower'	Very processive — 'porter'
Independence	Works in large groups	Works individually or in small groups
Working stroke	≈5–10 nm	≈8 nm/molecule (16 nm/head)
Path distance	≈36 nm	≈8 nm (16 nm/head)

Biological questions

Single-molecule mechanical experiments are still relatively new and to date most studies have concentrated on muscle myosin-II isoforms (e.g. smooth-, skeletal- and cardiac-muscle myosins), which are the prototypical system. The reader should see Chapter 3 in this volume for a description of the structure and function of myosin II and Chapter 4 in this volume for a discussion of the unconventional myosins. There are now known to be 15 different families of myosins (see Chapter 4 in this volume), and we expect that the biochemical kinetics and motor functions may be quite diverse both within and between these families.

Comparative studies of different myosins should yield insights into common mechanistic features and the biochemical and structural adaptations that tune the motor function for efficiency, speed, force or the economical maintenance of tension. For instance, we would like to know:
(i)   why some myosins have two heads and others only one;
(ii) if some myosins might be processive or even run 'backwards' (like kinesin; see Table 1);
(iii) if movement, force and stiffness are the same for all myosins;
(iv) how the power strokes are coupled to different biochemical steps;
(v)   if the ATPase cycle and mechanical cycle always have a tightly coupled, one-ATP-to-one-power stroke, relationship.

As new families and isoforms of myosins have been identified, motility assays *in vitro* (see below) have played an important role in demonstrating that many have motor activity (e.g. myosin I and myosin V). In the future, single-molecule mechanical and biochemical studies will enable us to discover how these myosins generate mechanical work from chemical 'fuel' (in the form of ATP).

Motility assays

Motility assays *in vitro* [2] enable a motile system to be reconstituted from its component molecules. The actomyosin motility assay was developed in the laboratories of Yanagida and Spudich in the 1980s. These assays have many variations. Typically, however, fluorescently labelled actin filaments slide over a microscope coverslip coated with myosin. This occurs in the presence of ATP. A sensitive video camera is used to record the motility, and the filaments are tracked by computer, allowing the speed of sliding to be determined. For example, rabbit skeletal-muscle myosin and actin produce a sliding velocity of around 5–9 μm·s^{-1}, similar to the maximum shortening velocity of intact muscle.

Assays *in vitro* reduce the system to its basic functional components and much has been learned from them because of the ease with which the reaction conditions can be altered, or different types of myosins, actins or regulatory proteins can be introduced. Some of the important issues that have been settled are:

(i)   purified actin and myosin and a buffered MgATP solution are all that is required for motility;

(ii) the polarity of the actin filament determines the direction of movement — the 'pointed end' of actin always leads;

(iii) the unloaded sliding velocity is close to that measured in intact muscle fibres, and is determined by the myosin isotype;

(iv) myosin subfragment-1 (containing the motor and regulatory domains) is sufficient for motility;

(v)   myosin heads can rotate through 180° and still support actin-filament sliding.

These experiments were also used to investigate the coupling of the biochemical and mechanical cycles of myosin, and began an important controversy that continues to this day. The question asked is: is the hydrolysis of one ATP molecule always tightly linked to the production of a single mechanical impulse? This is a key point, because it is central to any understanding of the mechanism by which myosin transmutes biochemical energy into mechanical work.

Tight coupling would mean that the answer to the question above is yes: the hydrolysis of one ATP always yields one mechanical step. *Loose coupling* would mean that the number of steps per ATP can vary in some way. The rotation of two gear wheels is tightly coupled, because the teeth cannot slip past one another. In contrast, a belt-drive system linking two smooth wheels allows some slippage. However, while such a system might tend to slip at high loads, thereby wasting energy, it has been suggested that myosin might instead make extra steps per ATP at low loads, and so save energy. To address this and other, more subtle, issues requires the study of single molecules.

Why single molecules?

Using bulk methods to investigate the kinetics of mechanical or biochemical steps in the actomyosin pathway requires that the starting states of the molecules be synchronized. This can be achieved by several specialized techniques: rapid mixing experiments, temperature or pressure jumps, step-length changes in muscle fibres or the use of 'caged compounds' (e.g. caged ATP) released by flash photolysis. We expect the myosin molecules in muscle fibres to behave in a heterogeneous fashion, because the spacing of binding sites on the actin filament means that some attached molecules will be under more strain than others; likewise, some detached molecules will be closer to a potential binding site than others. Therefore, it is unclear if the data obtained are (i) an average of *all* the myosins, (ii) only the pool of molecules that are bound to actin or (iii) just a small fraction that have some particular extreme property (e.g. high load or slow kinetics). This greatly complicates interpretation of results, particularly if there is rapid exchange between different pools of molecules.

We believe that the basic mechanism deduced from modelling of the average biochemical and mechanical behaviour is likely to be correct [3]. However, there are important subtleties that will be resolved only by studying the turnover of individual molecules directly. Furthermore, the mechanical properties of newly discovered cellular or 'unconventional' myosins can *only* be studied using single-molecule methods.

Molecular mechanics

At the scale of single molecules, 'life' is dominated by thermal energy [given by kT, where k is Boltzmann's constant (1.4×10^{-23} J·K^{-1}) and T is the absolute temperature in degrees Kelvin. At room temperature, the thermal energy is 4 pN·nm (here we use the mechanical units of pN·nm, since they are rather more intuitive than yoctojoules)]. At absolute zero, molecular vibrations cease and chemical reactions stop. But at room temperature, thermal energy does sufficient mechanical work to bend and distort molecules and to allow chemical reactions to proceed. During the myosin ATPase cycle, shape changes occurring at the catalytic site propagate across the molecule affecting, at one end, its affinity for actin and, at the other, the disposition of the regulatory domain (lever arm). If prevented from moving, the molecule becomes distorted and force is generated. By storing mechanical work as elastic strain energy, myosin captures the sudden (picosecond) changes in chemical potential associated with steps in the biochemical cycle, and does external work on a much slower timescale (milliseconds or seconds), e.g. as muscle shortens or vesicles are transported.

From muscle-fibre experiments, we know that myosin produces a force of 1–10 pN and that the movement will be 1–20 nm. Correspondingly, the stiffness will be a few $pN \cdot nm^{-1}$ or less. We also know that the total energy available must be less than the free energy of hydrolysis of one ATP molecule (<50 pN·nm). This is sufficiently low that the expected signals will be not much greater than thermal motion (see the definition of thermal energy above; $\approx$4 pN·nm); the signal-to-noise ratio is therefore unavoidably low and, most importantly, this thermal noise is *in the system being studied* rather than the instrumentation. Finally, because the attached lifetime of myosin is very brief ($\approx$5 ms), the molecules will diffuse apart and repetitive interactions will not be observed, unless actin and myosin are both held in position. This is in contrast to processive motors (e.g. kinesin; see Table 1 and Chapter 6 in this volume), which are more amenable to study, because they remain attached for many cycles.

Technologies

A *transducer* (a device that produces an electrical signal proportional to movement or force) of the correct sensitivity is required to measure the force and movement produced by a single myosin. Three approaches have been used for actomyosin studies: glass microneedles, optical tweezers and the atomic force microscope (AFM).

Microneedles

The first device to be built [4] consisted of an extremely fine glass needle that enabled mechanical recordings to be made from a small ensemble of myosin molecules interacting with a single actin filament (Figure 1a). These experiments showed that the elementary forces produced by the myosins were stochastic and independent of one another. They also showed that the average force produced by a single myosin head was about 1 pN and that the basic mechanism of force production during rapid-filament sliding was different from that when sliding was prevented and maximum force generated.

Optical tweezers

In 1994 the first measurements of force and movement produced by individual myosin molecules were made [5]. These experiments used an optical tweezer or optical trap. This device is constructed around a microscope, into which a powerful laser beam is introduced. The optics are arranged such that the beam is focused by the microscope's objective lens to the smallest possible diffraction-limited spot ($\approx$400 nm across). The steep gradients of light intensity around this point produce significant amounts of photon pressure, which 'pushes' high-refractive-index particles towards the focus [6]. The device can therefore be used to trap and manipulate micrometre-sized refractile

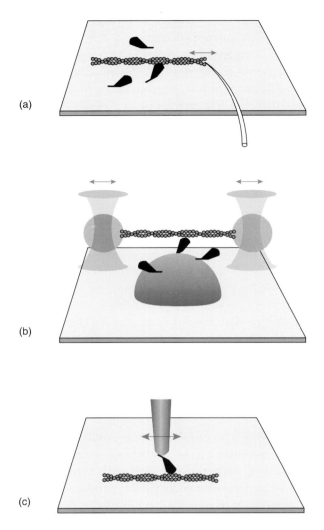

Figure 1. Techniques for single-molecule mechanical measurements
Three main techniques have been used to measure the mechanical behaviour of single motor molecules. (a) Glass microneedles: an actin filament is attached to a fine glass needle [4]. As the myosin interacts with actin, the deflection of the needle is monitored. (b) Optical tweezers: in the Finer et al. [5] 'three-bead' configuration an actin filament is suspended between two plastic beads in optical traps (shown in blue). Myosin motors (black shapes) on a third bead interact with the actin. The position of the beads in one or both traps is monitored (blue arrows). (c) Atomic force microscopy: myosin is attached to a scanning probe that is brought into contact with actin immobilized on a surface [7]. Deflections of the probe are monitored.

objects, such as cells or plastic beads — hence the name optical tweezer or trap. For spherical objects, the restoring force is linearly related to displacement from the centre of the trap, i.e. the optical trap behaves like a spring. Plastic beads are used as 'handles' to which actin or myosin may be attached.

The 'three-bead' arrangement (Figure 1b) devised by Finer et al. [5] is most commonly used. In this technique, an actin filament is held tensioned between two beads in independent traps, and the position of one of the beads is monitored by imaging it on to a quadrant photodiode. This detector measures the position of the bead to a resolution of better than 0.5 nm. The bead–actin–bead dumb-bell is held over a third bead on which is deposited a small number of myosin molecules, such that only a single molecule interacts at any one time. The rest of this chapter focuses on the use of this technique.

Table 2. Power-stroke measurements

A selection of results from single-molecule measurements of the myosin power stroke. The muscle myosins used are often in the form of proteolytic fragments. An indication is given of the fragment used: S1 (subfragment 1) consists of a single myosin motor and regulatory domain; HMM (heavy meromyosin) consists of two such domains linked by a coiled-coil region; single-headed myosin is a whole myosin molecule from which one of the heads has been removed by proteolysis. Some of the differences between these values may be due to interpretation or analysis of optical-trap data; others to the techniques used or to the properties of the different myosin types. Note that [14] presents evidence that the value may depend on the angle between actin and myosin filaments. Authors of [8,10–12] all agree that single-headed and double-headed myosin fragments give the same displacement, whereas the authors of [22] found that double-headed myosin gives twice the displacement. There is also some evidence for multiple displacement steps from one myosin: two steps for myosin I but not for S1 [11], and multiple 5.3 nm steps from S1 [7].

Type of myosin	Approximate power-stroke measured (nm)	Reference
Skeletal muscle (HMM)	11	[5]
Skeletal muscle (HMM and S1)	5	[8,11,12]
Skeletal muscle (HMM)	5	[10]
Skeletal muscle (single-headed)	0–15	[14]
Skeletal muscle	15	[19]
Skeletal muscle (S1)	11–30	[7]
Skeletal and smooth muscle (whole myosin)	13	[22]
Skeletal and smooth muscle (single-headed)	6–7	[22]
Myosin I	6.5 + 5.5 = 12	[11]

AFM

Recently, Kitamura et al. [7] developed an instrument based on the AFM. A single myosin molecule was attached to a very fine probe and then brought into contact with an actin-filament bundle and displacements and forces determined from movements of the probe tip (Figure 1c).

Displacement

Background

To determine the displacement, or working stroke, produced by a single actomyosin interaction, the stiffness of the apparatus must be much *less than* that of the actomyosin complex (so that myosin may undergo its full working stroke unhindered). To measure displacement, most workers use transducers with stiffness of ≈ 0.05 pN·nm^{-1}. At such low stiffness the transducer (either beads held in the optical tweezers, a glass microneedle or an AFM tip) necessarily exhibits large amounts of thermal motion. Interpretation of records obtained from this type of experiment is not straightforward, and different types of analysis give different estimates of the working stroke (see Table 2).

Experiments

Finer et al. [5] noted that attachment of a myosin head to the actin filament increases the stiffness of the link between the actin filament and 'mechanical ground', with a corresponding reduction in thermal movement of the bead–actin–bead assembly. This reduction in thermal noise was subsequently used as the criterion for identifying attachment [8] (Figure 2). By measuring the amplitude of each attachment and plotting a histogram, Molloy et al. [8] found that the breadth of the distribution of displacements was the same as that of the freely moving actin filament. By comparing the distribution of positions when the myosin was attached and unattached, they realized that the power stroke of the myosin could be deduced from the amount by which the attached distribution was shifted. For skeletal-muscle myosin, they found a displacement of 5 nm. Their analysis also explained the presence of very large steps and negative steps present in the data sets.

This interpretation has now become widely accepted, but a variety of different analysis methods are also used, including mean-variance analysis [9] and correlation between the two bead positions [10]. An extension of this approach is to use the change in stiffness to define the beginnings and ends of events, and then to average many together [11]. This 'ensemble averaging' gives information about the sequence of processes occurring. For example, it was shown recently that myosin-I isoforms have a two-step power stroke; this is consistent with structural data [11].

The time resolution for detecting the onset of myosin attachments (or detecting very brief interactions) is limited by the speed with which changes in thermal noise can be detected. Because the transducers are heavily damped, it is difficult to detect binding events that are of shorter duration than the relax-

(a)

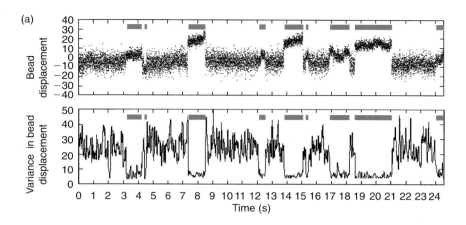

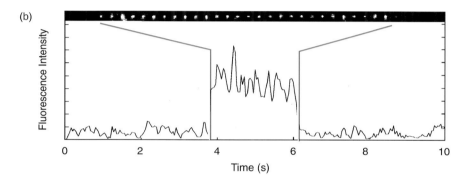

Figure 2. Traces from single-molecule experiments

(a) The upper panel shows a typical displacement trace from an optical-tweezers experiment. Note the high level of noise throughout the record (roughly ±10 nm root mean square), which is due to the thermal motion of the actin filament and beads. When a myosin (in this case a myosin I) binds to the actin filament, the noise level reduces because of the additional stiffness of the actomyosin link. This is illustrated by the lower panel, which shows the variance of the data. By using variance of the data to identify binding events (which are indicated by the black bars), we recover events with a wide distribution of displacements. This distribution arises because the starting positions of the displacements are random, but the end points are displaced through a fixed distance. The difference between the centres of the two distributions gives us the myosin's power stroke. (b) The plot shows 10 s of data from a single-molecule fluorescence experiment. Total internal-reflection fluorescence microscopy was used to visualize the binding of a fluorescent nucleotide analogue (10 nM Cy3-EDA-ATP) to a quartz surface sparsely coated with myosin I. The data were recorded using an image-intensified video camera; the strip at the top of the trace shows a sequence of images of one fluorescent spot. The trace shows the intensity of the spot over time (integrated over a 3×3 pixel area). Intensity rises and falls in a single step as nucleotide arrives and departs.

ation time constant (≈ 10 ms). Time resolution can be improved by applying a small amplitude, high-frequency oscillation to the transducer (e.g. applying a 1 kHz oscillation to the optical tweezer). Monitoring the amplitude of this oscillation enables changes in stiffness to be measured with approximately mil-

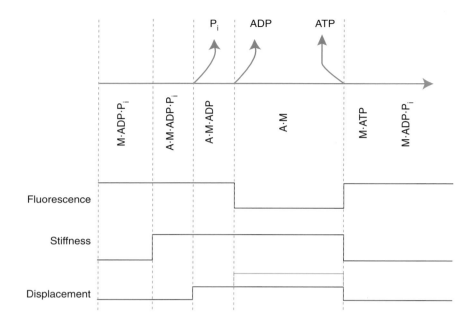

Figure 3. Schematic of the myosin mechanochemical cycle
This figure shows the presumed correlation between the biochemical events of the myosin ATPase cycle, the stiffness changes, displacement steps and fluorescence from nucleotide analogues. The stiffness increases when myosin attaches to actin. An initial displacement might arise as phosphate is released, which occurs very rapidly after myosin binds to actin. A second displacement step (blue line) has been observed for some myosin-I isoforms, and might also occur with smooth-muscle myosin II. This step may be coupled to the release of ADP. It is not clear whether skeletal-muscle myosins lack this step or whether the rapidity of ADP release in these myosins means that the two steps have not yet been resolved. The fluorescence signal corresponds to the presence of nucleotide in the active site; it should therefore decrease on ADP release and be restored on ATP binding. However, some results (see text) show that the fluorescence signal may drop before myosin binds to actin, which does not fit easily with the scheme shown here.

lisecond time resolution [12], which enables more rapid mechanical events to be detected.

Force

Accurate determinations of the force developed by myosin under *isometric* conditions (where there is no movement of the molecule) require that the stiffness of the transducer be much greater than that of the actomyosin complex (so that myosin is prevented from moving, allowing its maximum force to be developed). To measure force requires a transducer with a stiffness of ≈ 10 pN·nm^{-1} and the ability to measure movement with sub-nanometre precision. To achieve this, negative feedback is applied to the optical tweezer to compensate for bead movements measured by the position detector. When myosin binds and pulls on the actin filament, the force is counter-balanced by moving the optical tweezer in the opposite direction. Knowing the stiffness of the optical trap and the distance it has been moved, the force applied can be calculated. However, the situation is complicated because of compliance at the bead-to-actin connection [13]. This compliant linkage is stretched as myosin exerts force. Because the extension of this linkage is unknown, the maximum force that could be generated by myosin is underestimated (published values range from 1 to 5 pN).

This compliance can be corrected for by applying a large-amplitude, low-frequency oscillation to one of the optical tweezers, so that myosin is pushed and pulled several times while it is attached to actin. In this way, Veigel et al. [12] found myosin stiffness to be at least 0.7 pN·nm^{-1}.

Caveat: the orientation of actin and myosin

In addition to the problem of series compliance (mentioned above) there is a further issue affecting the interpretation of force and displacement measurements. In muscle, the interdigitating actin and myosin filaments have a fixed, parallel geometry. However, this is not the case in motility assays or in single-molecule experiments, where the myosin is randomly oriented on the surface (or on the AFM probe tip). The motor domain binds in a stereospecific manner to actin, and so must be aligned with the actin-filament axis. However, if the regulatory domain, or S2 (subfragment 2) part of the molecule is at some oblique angle then the working-stroke and maximum force may be reduced. Tanaka et al. [14] measured the angle dependence of the working stroke and found that over a wide range of angles (from 10 to 180° between myosin and actin) the displacement produced was not significantly different from 5 nm. However, when the angle between the myosin and actin was reduced to 6°, the working stroke increased to about 10±2 nm.

Kinetics

Background

Kinetic information can be extracted from single-molecule mechanical recordings by analysis of the distribution of attached lifetimes. Myosin on its own hydrolyses ATP only slowly, because release of phosphate from the catalytic site is slow. So the predominant *starting* form of myosin for each observed interaction will have both ADP and P_i bound in the catalytic site. Upon binding to actin there will be a series of biochemical transitions (see Figure 3 and Chapter 6 in this volume) as products are released and myosin (M) forms a tightly bound complex with actin (A) that generates both force and movement:

$$\cdots \longrightarrow A \cdot M \cdot ADP \cdot P_i \xrightarrow{\ 1\ } A \cdot M \cdot ADP \xrightarrow{\ 2\ } A \cdot M \cdots \longrightarrow \qquad (1)$$

Phosphate release (step 1) is rate-limiting in skeletal-muscle myosin, and is believed to be correlated with the power stroke. The release of ADP (step 2) then occurs fairly rapidly, leaving the myosin in a nucleotide-free or *rigor*

state. When a new molecule of ATP binds to myosin, it detaches rapidly from
the actin filament.

Experiments

To resolve the actomyosin bound state experimentally, most single-molecule
mechanical studies are performed at low ATP concentrations to artificially
prolong the rigor state. If the duration of a large number of attachments is
measured and a histogram plotted, the lifetimes follow a single exponential
decay (Figure 4), corresponding to first-order kinetics. By repeating the
experiment over a range of ATP concentrations, the kinetics of ATP binding
can be measured. As one might expect, kinetics determined in this way match
those obtained from bulk measurements.

Similar measurements can be made with other types of myosins, such as
myosin I. With some of the unconventional myosins, different steps in the
hydrolysis pathway may be rate limiting (e.g. ADP release or even ATP binding) and the kinetics are often slower. This may be helpful in dissecting the
mechanical and biochemical pathways. For example, an additional step is seen
in the displacement records obtained with brush-border myosin I and 130 kDa
myosin I. Veigel et al. [11] showed that the duration of the first phase of the
attachment is not affected by the concentration of ATP, whereas the second
phase is. This suggests that the additional power stroke corresponds to the
release of ADP or isomerization of an ADP-bound state.

Single-molecule fluorescence

The use of optical probes in combination with mechanical techniques should
give the detailed information that is required to understand the molecular basis
of force generation. Fluorescence is one of the most sensitive probes for
studying molecular behaviour but, because myosin, actin and ATP are only
weakly fluorescent, extrinsic high-intensity fluorescent probes must be
covalently attached to the molecules. This is done by chemical modification
with reactive dyes, or by expressing proteins as a fusion protein with green
fluorescent protein (GFP). Fluorescent ATP (e.g. Cy3-EDA-ATP) has been
found to be a suitable substrate for the myosin ATPase.

Detecting single-molecule fluorescence is difficult, but certainly not
impossible. The problems are as follows.

- *Sensitivity*: a single fluorophore produces an extremely small amount of
  light (a few thousand photons per second). Therefore it is important to
  maximize the number of photons collected (using high-numerical-aperture
  objectives and good-quality filters) and to use a very sensitive detector
  (such as a photomultiplier tube, avalanche photodiode or an intensified
  video camera).
- *Background*: this can be fluorescence from impurities in solutions, optics
  or slides; out-of-focus fluorescence from other fluorophores in the sample,

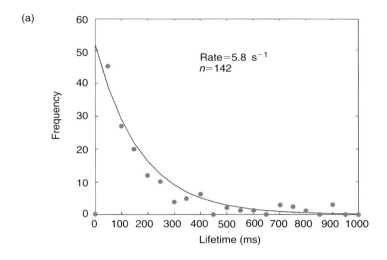

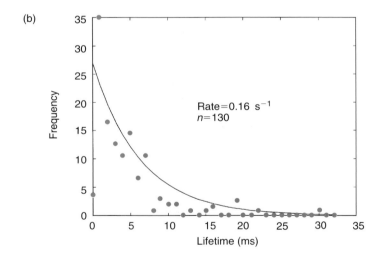

Figure 4. Measuring rates from single-molecule data

These graphs show how we can measure rate constants from single-molecule data. (a) An example of data from optical-tweezers experiments (see Figure 2a). The duration of the second phase of the attachment of Myr-1 (a myosin 1) was measured for 142 events in the presence of 10 μM ATP. The data points correspond to these lifetimes sorted into 50 ms bins. They show an exponential distribution, except that very few short events are observed, because it is more difficult to detect brief attachments. The curve corresponds to the calculated rate constant, accounting for missing data [11]. This rate constant measures the rate of ATP binding, and is dependent on the concentration of ATP. (b) Single-molecule fluorescence data from the same experiment as shown in Figure 2(b). For this myosin, the release of ADP is rate-limiting. Therefore, by measuring the lifetime of the fluorescent spots (upper strip in Figure 2b) we obtain a rate for ADP release, which should be independent of concentration.

or the excitation light 'leaking' through a filter. There are several ways of minimizing background; one of the most popular is a technique known as total internal-reflection fluorescence microscopy (TIRF) [15]. This technique only illuminates molecules within a very thin layer of solution near to the surface, and directs almost all of the unwanted excitation light out of the chamber.

- *Photobleaching*: when a fluorophore is raised into an excited state by an absorbed photon, there are several possible outcomes. The desired one is usually the emission of a photon of longer wavelength; this is what we mean by fluorescence. However, the excited molecule may also undergo an irreversible chemical reaction which renders it non-fluorescent. This process is called photobleaching, and must be minimized as far as possible. This is generally achieved by using the lowest possible excitation intensity and an enzymic system to remove oxygen from the experimental solution (an 'oxygen scavenger'). Under optimal conditions most fluorophores will emit an average of 10^7 photons before photobleaching (their lifetimes will be exponentially distributed). Knowing this, the illumination intensity and lifetime of the biochemical states can be adjusted to ensure that photobleaching events can be subtracted or otherwise cancelled out of the data set.

What kind of experiments can be done with myosin and actin using single-molecule fluorescence? The most basic experiment is to look at the kinetics of nucleotide binding and release [16]. In such experiments, myosin is bound at low surface density to a microscope coverslip and a very dilute solution (e.g. 1 nM) of Cy3-EDA-ATP is added. Using TIRF and a suitable detector, binding of individual ATP molecules and the subsequent release of ADP from myosins can be observed. Because small molecules move very rapidly in solution, they are not detected unless they become immobilized by binding to a protein molecule. An image of the diffraction-limited fluorescent spots produced by Cy3-EDA-ADP bound to myosin I is shown at the top of Figure 2(b). By measuring the lifetime of these spots, the kinetics of ADP release from myosin (Figure 4b) can be determined. For some myosins this step is rate-limiting and can be measured directly; for others the measured rate constant is a composite of earlier steps. If continuous observation of the same molecule is made, kinetics of ATP binding can be determined from the duration of the 'dark' intervals.

By observing the fluorescence of labelled proteins, other aspects of the actin-myosin system can be studied. For example, changes in the polarization of fluorophores attached to actin show that actin filaments rotate slowly as they slide over a myosin-coated surface [17]. Other researchers have used fluorescently labelled myosin to detect orientation changes of the regulatory domain during the power stroke [18].

Simultaneous studies

To understand mechanochemical coupling in myosin, we need to combine the above techniques to detect biochemical and mechanical events simultaneously in the same molecule. Two recent papers from the Yanagida laboratory describe this kind of combined measurement and explore different aspects of force generation in muscle myosins.

The first experiment combined optical tweezers and single-molecule fluorescence in one apparatus [19]. By comparing the timing of nucleotide binding and release, and the timing of mechanical events, they found that there is a 1:1 coupling between nucleotide-hydrolysis cycles and mechanical cycles. That is, one mechanical event is seen for each ATP consumed. They also found that, as expected, the binding of ATP causes an instantaneous release of myosin from the actin filament. More surprisingly, in some cases the nucleotide was released *before* actin binding and in these instances a power stroke was still produced if binding occurred within 2 s of nucleotide release. This means that the chemical energy obtained from ATP hydrolysis must somehow be stored in the detached myosin molecule. Perhaps there is a short-lived form of myosin that, while internally strained, remains competent to undergo a power stroke. Conformational changes that are induced by binding to actin would then propagate across the molecule, relieving internal strain and producing movement of the regulatory domain.

The other experiment combines single-molecule fluorescence with a type of AFM [7]. The fluorescent molecule was a labelled form of myosin subfragment 1. The fluorescence was used to confirm that only a single motor was attached to the end of the probe. When the probe was brought into contact with a bundle of actin filaments attached to a glass surface, multiple 5.3-nm steps were seen, giving total displacements of up to 30 nm. These results are surprising for several reasons: (i) they are different from what is seen with the optical tweezers; (ii) the total displacement seen is much higher than expected; and (iii) the multiple mechanical steps suggest some kind of loose coupling or energy storage between actin and myosin.

Conclusions

Figure 3 summarizes our current understanding of single-molecule observations made from actomyosin. The field is at an exciting stage; further experiments and development of new techniques will undoubtedly resolve the controversial issues and hopefully answer the question of how exactly muscles and myosins convert chemical energy into mechanical work.

Summary

- *Whereas we have a great deal of information about myosin, there remain fundamental questions about its mechanism (and those of other motor proteins).*

- *Single-molecule technologies enable us to make measurements we cannot make from large ensembles of molecules.*

- *Optical tweezers (and similar techniques) are used to measure the mechanical aspects of actomyosin interactions, including force, displacement and stiffness.*

- *Single-molecule fluorescence has been used to observe the binding and release of nucleotide by myosins.*

- *A combination of these measurements has the potential to solve the problem of coupling of ATP hydrolysis to mechanical work in motor proteins.*

The authors' work is supported by the BBSRC and the Royal Society. We thank John Sparrow and Claudia Veigel for their help with the manuscript.

References

1. Mehta, A.D., Rief, M., Spudich, J.A., Smith, D.A. & Simmons, R.M. (1999) Single-molecule biomechanics with optical methods. *Science* **283**, 1689–1695
2. Scholey, J.M. (ed.) (1993) Motility assays for motor proteins, *Methods in Cell Biology*, vol. 39 (Wilson, L. & Matsudaira, P., series eds.), Academic Press, San Diego
3. Cooke, R. (1997) Actomyosin interaction in striated muscle. *Physiol. Rev.* **77**, 671–697
4. Ishijima, A., Doi, T., Sakurada, K. & Yanagida, T. (1991) Sub-piconewton force fluctuations of actomyosin *in vitro. Nature (London)* **352**, 301–306
5. Finer, J.T., Simmons, R.M. & Spudich, J.A. (1994) Single myosin molecule mechanics: piconewton forces and nanometre steps. *Nature (London)* **368**, 113–119
6. Svoboda, K. & Block, S.M. (1994) Biological applications of optical forces. *Annu. Rev. Biophys. Biomol. Struct.* **23**, 247–85
7. Kitamura, K., Tokunaga, M., Iwane, A.H. & Yanagida, T. (1999) A single myosin head moves along an actin filament with regular steps of 5.3 nanometres. *Nature (London)* **397**, 129–134
8. Molloy, J.E., Burns, J.E., Kendrick-Jones, J., Tregear, R.T. & White, D.C.S. (1995) Movement and force produced by a single myosin head. *Nature (London)* **378**, 209–212
9. Guilford, W.H., Dupuis, D.E., Kennedy, G., Wu, J.R., Patlak, J.B. & Warshaw, D.M. (1997) Smooth muscle and skeletal muscle myosins produce similar unitary forces and displacements in the laser trap. *Biophys. J.* **72**, 1006–1021
10. Mehta, A.D., Finer, J.T. & Spudich, J.A. (1997) Detection of single-molecule interactions using correlated thermal diffusion. *Proc. Natl. Acad. Sci. U.S.A.* **94**, 7927–7931
11. Veigel, C., Coluccio, L.M., Jontes, J.D., Sparrow, J.C., Milligan, R.A. & Molloy, J.E. (1999) The motor protein myosin-I produces its working stroke in two steps. *Nature (London)* **398**, 530–533
12. Veigel, C., Bartoo, M.L., White, D.C.S., Sparrow, J.C. & Molloy, J.E. (1998) The stiffness of rabbit skeletal actomyosin cross-bridges determined with an optical tweezers transducer. *Biophys. J.* **75**, 1424–1438
13. Dupuis, D.E., Guilford, W.H., Wu, J. & Warshaw, D.M. (1997) Actin filament mechanics in the laser trap. *J. Muscle Res. Cell Motility* **18**, 17–30

14. Tanaka, H., Ishijima, A., Honda, M., Saito, K. & Yanagida, T. (1998) Orientation dependence of displacements by a single one-headed myosin relative to the actin filament. *Biophys. J.* **75**, 1886–1894

15. Axelrod, D. (1989) Total internal-reflection fluorescence microscopy. *Methods Cell Biol.* **30**, 245–270

16. Funatsu, T., Harada, Y., Tokunaga, M., Saito, K. & Yanagida, T. (1995) Imaging of single fluorescent molecules and individual ATP turnovers by single myosin molecules in aqueous-solution. *Nature (London)* **374**, 555–559

17. Sase, I., Miyata, H., Ishiwata, S. & Kinosita, K. (1997) Axial rotation of sliding actin filaments revealed by single-fluorophore imaging. *Proc. Natl. Acad. Sci. U.S.A.* **94**, 5646–5650

18. Warshaw, D.M., Hayes, E., Gaffney, D., Lauzon, A.M., Wu, J.R., Kennedy, G., Trybus, K., Lowey, S. & Berger, C. (1998) Myosin conformational states determined by single fluorophore polarization. *Proc. Natl. Acad. Sci. U.S.A.* **95**, 8034–8039

19. Ishijima, A., Kojima, H., Funatsu, T., Tokunaga, M., Higuchi, H., Tanaka, H. & Yanagida, T. (1998) Simultaneous observation of individual ATPase and mechanical events by a single myosin molecule during interaction with actin. *Cell* **92**, 161–171

20. Howard, J. (1997) Molecular motors: structural adaptations to cellular functions. *Nature (London)* **389**, 561–567

21. Leibler, S. & Huse, D.A. (1993) Porters versus rowers: a unified stochastic model of motor proteins. *J. Cell Biol.* **121**, 1357–1368

22. Tyska, M.J., Dupuis, D.E., Guilford, W.H., Patlak, J.B., Waller, G.S., Trybus, K.M., Warshaw, D.M. & Lowey, S. (1999) Two heads of myosin are better than one for generating force and motion. *Proc. Natl. Acad. Sci. U.S.A.* **96**, 4402–4407

6

Motor proteins of the kinesin superfamily: structure and mechanism

F. Jon Kull

Department of Biophysics, Max-Planck Institute for Medical Research, Jahnstrasse 29, 69120 Heidelberg, Germany

Introduction

Kinesins and kinesin-related proteins make up a large superfamily of molecular motors. The first kinesin was characterized by Vale et al. in 1985 as a cytosolic, microtubule-stimulated ATPase responsible for the directed, ATP-dependent movement of vesicles within squid axons [1]. Additionally, studies *in vitro* show that kinesin is able to translocate microtubules across a glass slide or latex beads along a microtubule, providing a powerful system for studying motor proteins.

Since their initial discovery, kinesins have been shown to be involved in movement of other cellular organelles and subcellular structures (e.g. mitochondria and chromosomes). The kinesins' primary role as an organelle transporter might at first seem surprising. Whereas the importance of muscle contraction is immediately apparent to anyone engaged in physical activity, the need for cellular components to be actively moved around inside cells may not be. Why not simply use diffusion? The answer is that diffusion, although quite rapid over small distances, becomes intolerably slow over larger distances. For example, whereas a small protein would take only 30 ms to diffuse across a 2 µm *Escherichia coli* cell, it would take over 2 h to diffuse along a 1 mm axon (and axons in human motor neurons can be up to 1 m long!). Therefore, kinesin's role as a cellular transporter is indeed essential for life to proceed at a reasonable speed.

The kinesin superfamily

The original 'conventional' kinesin was shown to be a tetrameric protein composed of two heavy chains (110–120 kDa) and two light chains (60–70 kDa). Electron microscopy, protease sensitivity and primary sequence analysis showed that the kinesin heavy chain is composed of three domains [2,3]. As shown in Figure 1(a), the globular N-terminal head domain (residues 1–325) contains the ATPase activity as well as a microtubule-binding site. The head is attached via a 50 amino acid neck region to an extended α-helical stalk (residues 375–800), which forms a coiled-coil upon dimerization with a second heavy chain. The C-terminal tail domain (residues 800–963) is globular and interacts with the kinesin light chains as well as with membrane-bound docking proteins such as kinectin [4].

Motility studies *in vitro* show that kinesin moves towards the plus-end of polar microtubule tracks (anterograde direction, away from the microtubule-organizing centre and toward the cell's periphery). Studies using single-molecule assays [5] have shown that conventional kinesin dimers move processively along a single microtubule protofilament, taking on average 100 8-nm steps per s from one α/β tubulin subunit to the next before dissociating [6]. Each step requires hydrolysis of one ATP molecule and produces a force of 5–7 pN [7].

Since the first kinesin was characterized, many related proteins have been discovered, including conventional kinesins as well as proteins belonging to the growing kinesin superfamily. Unlike the conventional kinesins, which share sequence identity throughout their entire sequences, these kinesin-related proteins share homology in only one region of ≈350–400 amino acids within the kinesin motor domain. Based on the location of the motor domain in the primary sequence, kinesin-family proteins fall into three classes, containing N-terminal, C-terminal or internal motors (Figure 1). The domain architecture of these proteins appears to be linked to their function. Whereas the N-terminal kinesins move exclusively towards the plus-ends of microtubules, all C-terminal motors characterized to date move towards the minus-end (retrograde, or toward the nucleus). Unlike kinesin, which can move processively along the microtubule without dissociating, the C-terminal motors are non-processive, disassociating from the microtubule following ATP hydrolysis. As more non-conventional kinesin motors have been described, other variations in chain composition have been identified, including functional homo- and heterodimers, trimers, tetramers and monomers (reviewed in [8]).

Molecular structure of the kinesin motor

The first X-ray structures of the motor domains of human conventional kinesin and Ncd (a C-terminal, minus-end motor from *Drosophila*) were solved by Kull et al. [9] and Sablin et al. [10] in 1996. Surprisingly, these structures of motors with opposite directionality were remarkably similar. Their ≈325 amino acid, arrowhead-shaped motor domains are composed of a

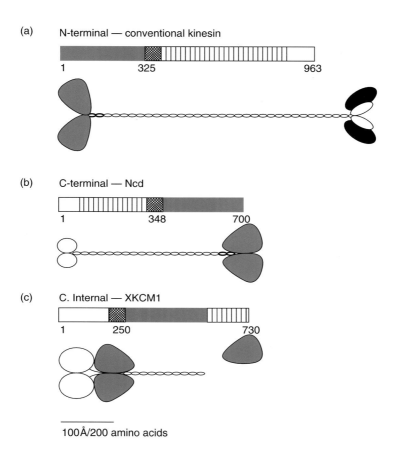

Figure 1. Schematic diagrams and amino acid domain structures of kinesin-family motor proteins of the N-terminal, C-terminal and internal motor types
(a) Human conventional kinesin; (b) *Drosophila* Ncd; and (c) *Xenopus* XKCM1. The core-motor head domain is indicated in blue, the subgroup-specific neck region by crosshatching (in the upper diagrams) and by thick black lines (in the lower schematic diagrams), the α-helical stalk region by vertical lines, and the tail domain in white. Kinesin light chains are in solid black. All molecules are drawn approximately to scale and depicted as dimers (scale bar, 100 Å; 200 amino acid residues).

core eight-stranded, mostly parallel, β-sheet flanked on each side by three α-helices. The MgADP seen in both structures is bound in a rather open, solvent-exposed cleft (Figure 2).

Although kinesin contains a phosphate-binding loop (P-loop) virtually identical to those found in other nucleotide-binding proteins (such as adenylate kinase, transducin and recA), there is little similarity in other regions. In contrast, structural elements of kinesin and myosin virtually superpose with one another in the core region surrounding the nucleotide-binding site. In all, 250 amino acid residues, including seven core β-strands and six α-helices, overlap. This degree of structural similarity is surprising because myosin's

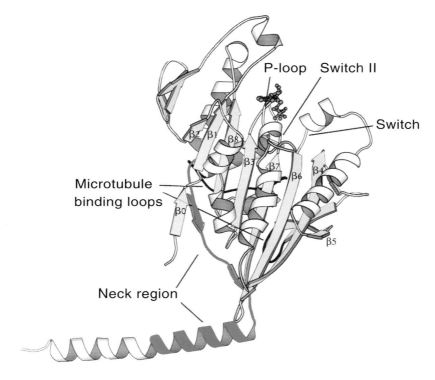

Figure 2. Structure of monomeric conventional kinesin from rat (PDB-2KIN) [25]
Loops and α-helices are shown in white, the kinesin-specific neck region in dark blue, β-strands in light blue, core β-strands are numbered. The bound ADP nucleotide is shown as a grey ball-and-stick figure. The nucleotide-binding motifs common to kinesin, myosin and GTPases are indicated (P-loop, switch I and switch II). The two polymer-specific insertion regions common to kinesin and myosin are indicated as black loops on the back aspect of the protein.

motor domain is much larger than kinesin's and there is very little sequence similarity between the two proteins. The Ras family of GTPases also share a common core nucleotide-binding region with kinesin, but with less structural overlap (six β-strands and four α-helices). The most striking similarities are seen in the regions directly adjacent to the ADP-binding site, including the P-loop and two other highly conserved motifs, called switch I and switch II, which sense the state of the bound nucleotide and 'switch' conformation as GTP is hydrolysed to GDP [11].

The structural elements shared between kinesin and myosin do not compose a single, continuous domain. Instead, the motors contain additional domains inserted between their shared elements at two points that might be involved in specific interactions with their respective polymer tracks. In kinesin, these regions consist of two short loops involved in microtubule binding (Figure 2), whereas in myosin these loops are replaced by large actin-binding regions of 150–200 amino acids. Thus it seems that the two motor families use a common core with different domain insertions to confer polymer specificity. These observations, in combination with highly conserved side-chain positioning and chemistry in the active site, imply that kinesins and myosins have evolved from an ancestral motor protein. An evolutionary connection to G-proteins is more tenuous; however, the shared structural, functional and chemical features of motors and these GTPases hint that an ancestral nucleotide-binding protein could have diverged to become both an ATP-driven motor protein and a GTP-driven molecular switch.

The initial crystal structures of kinesin and Ncd posed more questions than they answered. They showed neither a mobile lever-arm structure, as seen in the myosin motor (see Chapter 3 in this volume), nor an obvious structural difference that could explain their opposite directionalities. Fortunately, the recent crystal structures of dimeric rat conventional kinesin by Kozielski et al. [12] and of dimeric Ncd by Sablin et al. [13], in combination with primary sequence analysis and the construction of chimaeric kinesin motors [14–16], have clarified some of these questions. The neck region of these proteins is the primary structural element determining directionality, and conventional kinesin produces force via a 'hand-over-hand' mechanism, essentially walking along a microtubule in a manner quite unlike myosin.

The crystal structure of dimeric conventional kinesin shows the 379 amino acid construct held together via a 30 amino acid coiled-coil interaction in the neck region. Interestingly, the 2-fold rotational symmetry of the coiled-coil is not adopted by the motor heads, which orient themselves in a tip-to-tip manner (Figure 3a). In addition to the helix, the kinesin dimer contains a neck linker region that forms two additional β-strands, present but unordered in the original monomeric crystal structure (see Figures 2 and 3a).

In the Ncd dimer structure (Figure 3b), the heads are once again connected via a coiled-coil neck helix, and show 2-fold rotational symmetry. The 30 amino acid neck linker of the kinesin dimer is replaced by a very short turn resulting in a more compact dimer containing many more interactions between the heads and the coiled-coil.

Further structural data contributing to a working hypothesis for the structural basis of kinesin motilty come from three-dimensional electron-microscopic (EM) reconstructions of microtubules decorated with motor heads. Although initial experiments with monomeric heads showed little difference between kinesin and Ncd, subsequent experiments with dimeric heads have yielded interesting results [17,18]. As depicted in Figure 4, dimeric kinesin containing the non-hydrolysable nucleotide analogue AMP-PNP (adenosine 5′-[β,γ-imido]triphosphate) shows one head clearly bound to the microtubule with the second head oriented above the first, off the microtubule surface, and towards its plus-end. A similar orientation of heads could be obtained when the dimeric crystal structure is modelled into the EM density with its putative microtubule-binding face down (compare Figures 3a and 4a). In contrast, the unbound head of dimeric Ncd is located directly above and to the side of the bound head, with no positioning towards the plus-end.

A. Kinesin

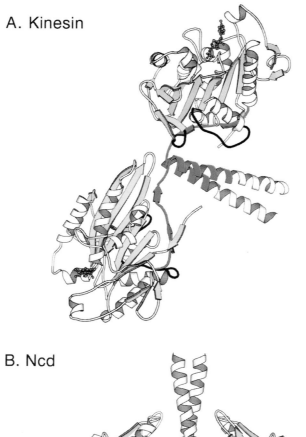

B. Ncd

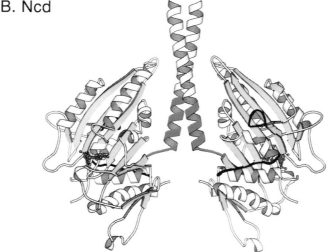

Figure 3. Structures of dimeric kinesin family of proteins
(a) Rat conventional kinesin dimer (PDB entry 3KIN). (b) *Drosophila* Ncd (PDB entry 2NCD). Structural elements are coloured as in Figure 2. If these structures were docked to a microtubule, the bound head (left in both structures) would be rotated ≈180° from the orientation shown in Figure 2. Note the unbound head of kinesin is oriented towards the top of the page (the plus-end of the microtubule), whereas the unbound head of Ncd is oriented closer to the bottom of the page (minus-end of the microtubule). See also the diagrams on the right-hand side of Figure 4.

Recent analysis of the conformation of dimeric heads of kinesin and Ncd in different nucleotide states has shown that the unbound head of kinesin·ADP is not oriented towards the microtubule plus end (Figure 4) [19,20]. The heads assume this conformation only in the presence of AMP-PNP (and also, perhaps, in the ADP·P$_i$ state). The experiments indicate that the orientation of the bound motor heads on the microtubule is responsible for directionality. In conventional kinesin, following binding of a non-hydrolysable ATP analogue, the unbound head is positioned much closer to the next available plus-end-binding site than to the minus-end site. For Ncd, the situation is less clear because the unbound head is not directed towards either the plus- or minus-end and does not change orientation appreciably between different nucleotide states.

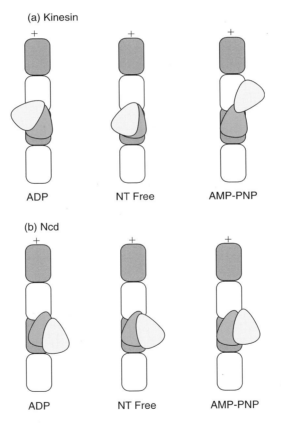

(a) Kinesin

ADP   NT Free   AMP-PNP

(b) Ncd

ADP   NT Free   AMP-PNP

Figure 4. Schematic diagram of kinesin– and Ncd–microtubule complexes as derived from EM three-dimensional reconstructions [19,20]

The relative positions of the heads of the kinesin dimer (a) and Ncd dimer (b) are shown in three different nucleotide states. The bound head is dark blue, the unbound head light blue, the α- and β-tubulin subunits are white and grey, respectively. + indicates the plus-end of the microtubule. NT, nucleotide.

The mechanochemical mechanism of conventional kinesin

It is only in the last year that enough structural, kinetic and functional data have accumulated to suggest a mechanochemical mechanism for the movement of conventional kinesin along a microtubule (Figure 5). The model starts with a kinesin dimer in solution with ADP bound to both heads. Kinetic studies [21] indicate that binding of the dimer (step 0) to the microtubule results in rapid loss of ADP from the bound head (step 1). This is accompanied by a transition from a weak-binding to a strong-binding state of the bound head, and a small movement of the unbound head. ATP binding triggers a rotational movement of the second, unbound head, resulting in a plus-end-biased orientation (step 2). The unbound head searches via a diffusional mechanism for an available microtubule-binding site, binds (step 3), quickly releases ADP and adopts a tight binding conformation. The first head, now lagging, hydrolyses ATP, undergoes a conformational change upon ADP or P_i release and disassociates from the microtubule. This restores the step-1 conformation with the heads reversed (step 1'). Kinesin is designed such that the two heads communicate with each other through ATP-dependent conformational changes, in which binding of the free head to the microtubule can only occur following ATP binding in the lagging head. In this manner, dimeric kinesin can walk processively and unidirectionally along a microtubule.

Variations on a theme: walkers, pushers, sliders and destabilizers

It should be emphasized that many details of this mechanism are poorly understood on a structural level. Furthermore, there are at least three other methods by which kinesin family members seem to operate: (i) pushing, as represented by the C-terminal minus-end-directed motors Ncd and Kar3; (ii) sliding, used by the processive monomeric motor KIF1A; and (iii) destabilizing, as used by the microtubule-destabilizing kinesin XKCM1.

Despite extensive structural, kinetic and mutational studies on the minus-end-directed motor Ncd, and the recent determination by Gulick et al. [22] of the crystal structure of Kar3 (another minus-end-directed motor), a plausible mechanism for reversed directionality remains elusive. The 2-fold rotational symmetry of Ncd's heads seems designed to keep the motor domains together rather than allowing them to spread apart. Chimaeric and deletion constructs show that Ncd motors lacking the neck region become poor plus-end-directed motors (similar to monomeric deletion constructs of conventional kinesin). This indicates an inherent plus-end-directed movement in the core motor of kinesin proteins [14–16]. Conversely, placing Ncd's neck region on the kinesin motor domain forces minus-end-directed movement for the normally plus-end-directed motor. Both of these findings indicate that the neck region of Ncd is necessary for producing minus-end-directed movement.

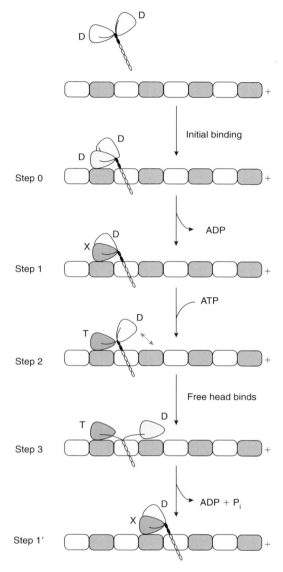

Figure 5. The hand-over-hand mechanism for the mechanochemical cycle of conventional kinesin

The bound nucleotides are indicated by D (ADP) and T (ATP); X is nucleotide free. Unbound heads are white, weakly bound heads light blue, and tightly bound heads dark blue. The cycle is initiated by dimeric kinesin binding to the microtubule (step 0). ADP is then released quickly from the bound head, accompanied by a transition from a weak to a strong binding state and a slight conformational shift of the unbound head (step 1). ATP binding then induces a major conformational change, positioning the free head towards the next plus-end microtubule-binding site (step 2). This head then finds a binding site via a biased diffusional search (blue arrow), accompanied by a melting/unwinding of the neck region (thick black lines; step 3). The first head now hydrolyses its ATP, and (in unknown order) releases from the microtubule and releases ADP and P_i, restoring the configuration of the initial state with heads now reversed (step 1').

Ncd could employ the same general mechanism as kinesin, in which the position of the unbound head influences the next microtubule-binding site. For dimeric Ncd, the microtubule-binding site closest to the unbound head would be on the adjacent protofilament (Figure 6a). In order to reach this site, the neck helix of Ncd would have to melt, as it seems capable of doing [13]. By repeatedly binding and releasing in such a manner, multiple Ncd dimers could produce non-processive, minus-end movement.

Another variation on the kinesin motility theme was discovered recently by Okada and Hirokawa [23], where they found that the monomeric, N-terminal kinesin motor KIF1A is able to move processively along a microtubule track. This result is extraordinary given that monomeric constructs of conventional kinesin have been shown to be non-processive, requiring multiple motors in order to produce movement *in vitro*. Apparently KIF1A employs a positively charged lysine tether to bind electrostatically to the negatively charged microtubule while it moves to the next binding site. The inherent plus-end movement of the kinesin motor domain, described above, biases the sliding diffusion of KIF1A to the plus-end, resulting in net movement in that direction.

One final twist of kinesin's mechanism involves two members of the internal motor-domain kinesins, XKCM1 and XKIF2. Desai et al. [24] recently showed that these non-motile kinesins are actually microtubule-depolymerizing factors that act *in vitro* by binding to the ends of microtubules and inducing catastrophe (rapid microtubule depolymerization). The destabilizing activity is dependent on ATP hydrolysis, which seems to be necessary for the release of α/β tubulin from the kinesin and permits its subsequent reattachment to the microtubule's end. Desai et al. have speculated that regulation of polymer dynamics most probably preceeded motor-protein-based motility in cells. If so, this family of kinesins might represent a missing link between a primitive molecular switch and a modern molecular motor.

Perspectives

The model of how conventional kinesin converts chemical energy into directed force must now be improved to explain the detailed conformational changes that accompany kinesin's hand-over-hand mechanism. Although the existing crystal structures have been invaluable, they are still only a static picture of the ADP nucleotide state. Electron microsopy has helped to identify conformations present in other nucleotide states, but at much lower resolution. In order to understand this system more fully, it will be necessary to obtain atomic-level structures of kinesin motors in other nucleotide states, as well as in a complex with microtubules.

Many other questions remain unanswered. The neck region of kinesin seems critical for controlling the motor's directionality, but the details of this are unknown. While a hand-over-hand mechanism seems certain for conven-

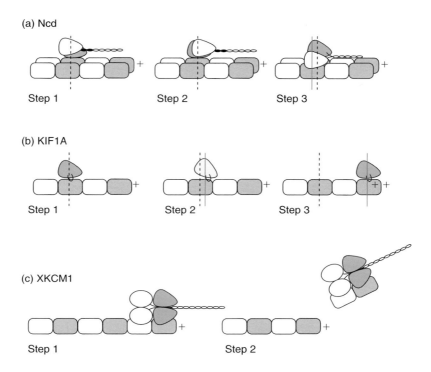

Figure 6. Hypothetical mechanisms of other kinesin-family proteins
(a) Model for the non-processive movement of dimeric Ncd to the minus end of the micro-tubule. Step 1 shows Ncd dimer bound to the microtubule. Binding and hydrolysis of ATP results in a conformational change of the unbound head (step 2) and its subsequent attachment to a binding site on an adjacent microtubule protofilament (step 3). Release of the forward head, cou-pled to repetition of this by multiple Ncd motors would result in a net displacement to the minus-end. (b) Model for the processive movement of KIF1A along a microtubule [23]. KIF1A contains a modified microtubule-binding loop containing a poly-lysine repeat. This charge tether allows the protein to remain associated with the microtubule while diffusing along it to the next binding site. The inherent plus-end-directed conformational change of the kinesin head as it hydrolyses ATP (see text) biases this diffusion in the plus-end direction. (c) Model for the disrup-tion of microtubule protofilament ends by the XKCM1 kinesin [24]. XKCM1 binds to the an α/β tubulin dimer at the plus-end of a microtubule protofilament (step 1) and causes dissociation of the XKCM1–α/β tubulin complex (step 2). ATP hydrolysis releases the XKCM1 from the α/β dimer and primes it for reattachment to the end of the microtubule.

tional kinesins, how do the non-processive motors produce force? How is kinesin activity regulated? What roles do docking proteins such as kinectin play? Although the past several years have seen great advances, recent discov-eries of a processive monomeric kinesin and kinesins that do not seem to be motors at all indicate that we still have far to go to fully understand the many cellular functions and mechanisms of this diverse protein family.

Summary

- *Kinesins are ATP-driven microtubule motor proteins that produce directed force.*
- *The kinesin superfamily currently encompasses over 100 eukaryotic proteins containing a common motor domain. Both the nucleotide-binding fold and active-site chemistry of the motor domain are also present in the actin-based motor, myosin.*
- *Kinesins can be classified into three groups based on the position of their motor domains: N-terminal, C-terminal and internal kinesins.*
- *Conventional kinesin operates as a dimer, walking in a co-ordinated, hand-over-hand fashion along a microtubule protofilament.*
- *X-ray crystal structures and EM reconstructions show major differences in the quaternary arrangement of kinesin domains in minus-end- and plus-end-directed motors.*
- *Kinesin's neck region, directly adjacent to the motor domain, dictates directionality.*

I am grateful to R. Batra, M. Knetsch, D. Manstein and F. Kull for comments and critical reading of the manuscript; R. Fletterick for allowing me to use the Ncd co-ordinates not yet available from the Protein Data Bank; L. Amos and R. Cross for insight and permission to discuss unpublished results; and the Max-Planck Gesellschaft and Deutsche Forschungsgemeinschaft for research support. Explore the kinesin home page at: http://www.proweb.org/kinesin.

References

1. Vale, R.D., Reese, T.S. & Sheetz, M.P. (1985) Identification of a novel force-generating protein, kinesin, involved in microtubule-based motility. *Cell* **42**, 39–50
2. Bloom, G.S., Wagner, M.C., Pfister, K.K. & Brady, S.T. (1988) Native structure and physical properties of bovine brain kinesin and identification of the ATP-binding subunit polypeptide. *Biochemistry* **27**, 3409–3416
3. Yang, J.T., Laymon, R.A. & Goldstein, L.S. (1989) A three-domain structure of kinesin heavy chain revealed by DNA sequence and microtubule binding analyses. *Cell* **56**, 879–889
4. Sheetz, M.P. (1999) Motor and cargo interactions. *Eur. J. Biochem.* **262**, 19–25
5. Vale, R.D., Funatsu, T., Pierce, D.W., Romberg, L., Harada, Y. & Yanagida, T. (1996) Direct observation of single kinesin molecules moving along microtubules. *Nature (London)* **380**, 451–453
6. Schnitzer, M.J. & Block, S.M. (1997) Kinesin hydrolyses one ATP per 8-nm step. *Nature (London)* **388**, 386–390
7. Kojima, H., Muto, E., Higuchi, H. & Yanagida, T. (1997) Mechanics of single kinesin molecules measured by optical trapping nanometry. *Biophys. J.* **73**, 2012–2022
8. Vale, R.D. & Fletterick, R.J. (1997) The design plan of kinesin motors. *Annu. Rev. Cell Dev. Biol.* **13**, 745–777
9. Kull, F.J., Sablin, E.P., Lau, R., Fletterick, R.J. & Vale, R.D. (1996) Crystal structure of the kinesin motor domain reveals a structural similarity to myosin. *Nature (London)* **380**, 550–555
10. Sablin, E.P., Kull, F.J., Cooke, R., Vale, R.D. & Fletterick, R.J. (1996) Crystal structure of the motor domain of the kinesin-related motor Ncd. *Nature (London)* **380**, 555–559

11. Vale, R.D. (1996) Switches, latches, and amplifiers: common themes of G proteins and molecular motors. *J. Cell Biol.* **135**, 291–302

12. Kozielski, F., Sack, S., Marx, A., Thormahlen, M., Schonbrunn, E., Biou, V., Thompson, A., Mandelkow, E.M. & Mandelkow, E. (1997) The crystal structure of dimeric kinesin and implications for microtubule-dependent motility. *Cell* **91**, 985–994

13. Sablin, E.P., Case, R.B., Dai, S.C., Hart, C.L., Ruby, A., Vale, R.D. & Fletterick, R.J. (1998) Direction determination in the minus-end-directed kinesin motor ncd. *Nature (London)* **395**, 813–816

14. Henningsen, U. & Schliwa, M. (1997) Reversal in the direction of movement of a molecular motor. *Nature (London)* **389**, 93–96

15. Case, R.B., Pierce, D.W., Hom-Booher, N., Hart, C.L. & Vale, R.D. (1997) The directional preference of kinesin motors is specified by an element outside of the motor catalytic domain. *Cell* **90**, 959–966

16. Endow, S.A. & Waligora, K.W. (1998) Determinants of kinesin motor polarity. *Science* **281**, 1200–1202

17. Hirose, K., Lockhart, A., Cross, R.A. & Amos, L.A. (1996) Three-dimensional cryoelectron microscopy of dimeric kinesin and ncd motor domains on microtubules. *Proc. Natl. Acad. Sci. U.S.A.* **93**, 9539–9544

18. Hoenger, A., Sack, S., Thormahlen, M., Marx, A., Muller, J., Gross, H. & Mandelkow, E. (1998) Image reconstructions of microtubules decorated with monomeric and dimeric kinesins: comparison with X-ray structure and implications for motility. *J. Cell Biol.* **141**, 419–430

19. Arnal, I. & Wade, R.H. (1998) Nucleotide-dependent conformations of the kinesin dimer interacting with microtubules. *Structure* **6**, 33–38

20. Hirose, K., Löwe, J., Alonso, M., Cross, R.A. & Amos, L.A. (1999) Congruent docking of dimeric kinesin and Ncd into 3-D electron cryo-microscopy maps of microtubule-motor.ADP complexes. *Mol. Biol. Cell* **10**, 2063–2074

21. Ma, Y.Z. & Taylor, E.W. (1997) Interacting head mechanism of microtubule-kinesin ATPase. *J. Biol. Chem.* **272**, 724–730

22. Gulick, A.M., Song, H., Endow, S.A. & Rayment, I. (1998) X-ray crystal structure of the yeast Kar3 motor domain complexed with Mg.ADP to 2.3 Å resolution. *Biochemistry* **37**, 1769–1776

23. Okada, Y. & Hirokawa, N. (1999) A processive single-headed motor: kinesin superfamily protein KIF1A. *Science* **283**, 1152–1157

24. Desai, A., Verma, S., Mitchison, T.J. & Walczak, C.E. (1999) Kin I kinesins are microtubule-destabilizing enzymes. *Cell* **96**, 69–78

25. Sack, S., Muller, J., Marx, A., Thormahlen, M., Mandelkow, E.M., Brady, S.T. & Mandelkow, E. (1997) X-ray structure of motor and neck domains from rat brain kinesin. *Biochemistry* **36**, 16155–16165

7

The molecular anatomy of dynein

Alistair Harrison & Stephen M. King[1]

Department of Biochemistry, University of Connecticut Health Center, 263 Farmington Avenue, Farmington, CT 06032–3305, U.S.A.

Introduction

Dyneins act as molecular motors using energy derived from the hydrolysis of ATP to translocate towards the minus-end of microtubules. From the initial identification of dynein as an ATPase in *Tetrahymena* cilia and sea urchin sperm flagella, these enzymes and their components are now known or suggested to be involved in such varied processes as mitosis, left–right asymmetry during mammalian development, sperm development, the movement of membranous organelles, rhodopsin trafficking and intraflagellar transport. Classically, dyneins can be divided into three major classes dependent primarily on subcellular localization: these are the dyneins of the inner and outer arms of flagella, and cytoplasmic dynein. All of these enzymes are involved in the generation of force relative to a microtubule (Figure 1). Inner- and outer-arm dyneins generate complex flagellar waveforms by producing differential lateral forces between adjacent microtubules. The outer arms transduce a substantial proportion of the total force, while the inner arms are responsible for the fine control of the waveform. Conversely, cytoplasmic dynein has a much more varied role. It is involved in the movement of organelles and polarity determinants as well as retrograde transport in axons and maintenance of the Golgi apparatus. Cytoplasmic dynein also cross-links

[1]*To whom correspondence should be addressed.*

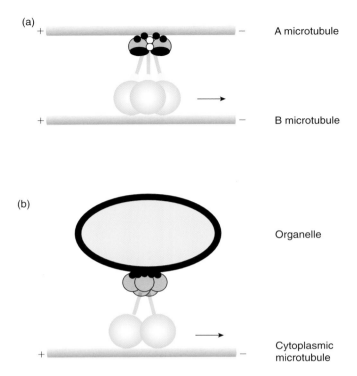

Figure 1. Physiological activities of dynein
(a) Flagellar inner- and outer-arm dyneins. These dyneins are attached to the A tubule of the outer doublet microtubules via an ATP-insensitive interaction that, at least for outer-arm dynein, is mediated by the intermediate chains and a newly identified docking complex. The globular head domains of the particle interact with the B tubule of an adjacent doublet and hydrolyse ATP to generate motive force. This results in a minus-end-directed sliding movement of one micro-tubule with respect to the other. Within the flagellum, the radial spokes convert this sliding motion into a bend, thereby generating a flagellar beat. (b) Cytoplasmic dynein binds various intracellular cargoes in an ATP-insensitive manner. Attachment to vesicles requires an adaptor unit known as dynactin that interacts with the basal intermediate-chain–light-chain complex. The motor head domains hydrolyse ATP and cause movement towards the microtubule minus-end (arrow).

and positions microtubules of the mitotic spindle. Furthermore, recent studies support a role for cytoplasmic dynein in sperm morphogenesis and intraflagellar transport.

Dyneins are massive multimeric enzymes with molecular masses between 1 and 2 MDa. The multiplicity of known dynein functions is mirrored by a corresponding structural complexity. They are composed of a number of heavy, intermediate and light polypeptide chains, the arrangement of which determines the identity and functioning of the dynein in question. Therefore, to understand how a particular dynein enzyme acts to transport the correct cargo to the appropriate location, it is essential to understand the underlying structural organization of the motor complex.

Force generation and microtubule movement

Dynein-mediated motility is powered by the hydrolysis of ATP. During the mechanochemical cycle, the enzyme binds ATP, hydrolyses it to ADP and phosphate, and subsequently releases the products. Conformational changes in protein structure are intimately associated with each step of the ATPase cycle and allow the motor enzyme to interact transiently with the microtubule to generate force. Following the force-generating step, the dynein must release from the microtubule in order to return to the original conformation and thus allow for the next cycle. In this manner the dynein motor is able to translocate along microtubules at rates of 1–6 $\mu m \cdot s^{-1}$ (the velocity depends on the source of dynein, buffer conditions etc.).

The structure of dynein

The basic structure of a generic dynein is illustrated in Figure 2. However, it is important to keep in mind that some dyneins diverge significantly from this overall plan. These enzymes are constructed around the massive ($\approx$500 kDa) dynein heavy chains, which form the globular motor domains, and the stems. The number of heavy chains can be from one to three depending on the dynein in question (see below). It is these components that hydrolyse ATP and interact with microtubules to generate force. At the base of the stems are located several intermediate chains that are involved in the attachment of the dynein to its particular cargo. Additional components are also present, including light chains associated with the stems (and in one case the head) of the heavy chains. Two classes of light chain (LC8- and Tctex1-family proteins) are found at the base as part of an intermediate-chain–light-chain complex.

Specific features of different dynein classes

Inner-arm dynein

In the alga *Chlamydomonas*, the inner flagellar arms (I1, I2 and I3) are arranged into groups of three that repeat every 96 nm. The inner arms are responsible for the initiation and propagation of flagellar bends. Inner arm I1 is located proximal to the first spoke of the radial spoke pair. It contains two heavy chains, 1α and 1β, at least two intermediate chains as well as Tctex1 and LC8. The groups of inner arms termed I2 and I3 are distinct from I1 and consist of single heavy chains associated with several other components, including p28, actin and centrin (there are multiple different members of these inner-arm groups). Moreover, these dyneins have a complex localization, with some being present in the region of the flagellum proximal to the cell body while others are assembled more distally.

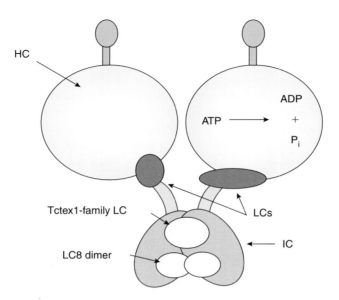

Figure 2. A generic model for the structure of dynein
A simplified model of dynein structure showing common features. Each heavy chain (HC) forms
a globular head (which hydrolyses ATP to produce motive force) and a stalk domain. Attached to
the heavy-chain stems are several light chains (LCs) and intermediate chains (ICs). The basal
intermediate-chain–light-chain complex is involved in cargo and/or adaptor binding.

Outer-arm dynein

Unlike the inner arms, there is only a single type of outer arm within a
flagellum. The outer arms contain either two or three different heavy chains
depending on source. These each have light chains bound and also interact
with an intermediate-chain–light-chain complex at the base. Mutational
analysis has revealed that the role of the outer arms is to provide the power for
flagellar beating. Lack of these components reduces beat frequency by ≈50%
but has little effect on the waveform itself. The outer arm interacts *in situ* with
an additional trimeric structure known as the docking complex. This complex
is essential for binding of the outer arm to the axoneme and appears to
determine the attachment site.

Cytoplasmic dynein

Unlike flagellar dyneins, the major cytoplasmic isozyme (known as 1a)
contains two apparently identical heavy chains, as well as an intermediate-
chain–light-chain complex analogous to that found in the outer arm. This basal
structure is involved in attachment to dynactin, which is essentially an adaptor
to mediate dynein–vesicle interactions. In addition, this complex also includes
four light intermediate chains of ≈53–59 kDa, but it is not yet clear where they
are located. Recently, a second class of cytoplasmic dynein (termed 1b) has

been recognized. This enzyme appears to consist of a single heavy chain; no accessory proteins have yet been described.

The heavy chains

Dyneins are constructed around large (>500 kDa) heavy chains that contain the ATPase motor domain of the enzyme. Many organisms express a number of different heavy chains: for example, *Chlamydomonas* has 16 such chains, two cytoplasmic and 14 flagellar-specific (M.E. Porter, personal communication). All of these proteins have a number of common features (Figure 3), including four centrally located nucleotide-binding consensus sequences (P-loops) that are separated by ≈350 amino acid residues [1]. P1, the P-loop (or phosphate-binding loop) nearest to the N-terminus is absolutely conserved in heavy chains from both cytoplasmic and axonemal dynein [1,2], whereas the other three loops are more diverse. Interestingly, the P3 and P4 loops are more highly conserved among cytoplasmic and axonemal dyneins, respectively, although the functional significance of this is unclear. The absolute conservation of the P1 loop between all heavy chains suggests functional importance. Indeed, this loop forms part of the ATP-hydrolysing site of the dynein heavy chain; disrupting the P1 loop leads to a complete loss of ATPase activity. The other P-loops still lack a clearly assigned function, although several lines of evidence suggest that they bind nucleotides and may therefore be involved in dynein regulation [3].

Co-operative interactions between heavy chains also occur. For example, in *Chlamydomonas* outer-arm dynein, the ATPase activity of the β heavy chain is down-regulated by association with the α heavy chain. Furthermore, the enzymic properties of the resulting αβ dimer are modified further by interaction with the γ heavy chain. This suggests that the overall activity of the outer arm, under different conditions, is determined by intercommunication between the heavy chains. Genetic and structural analyses have revealed that the motor domains of the three outer-arm heavy chains are arranged with the γ chain innermost and the α chain to the outside of the axoneme (Figure 4) [4]. The apparent direct contact between heavy chains could allow conformational changes in adjacent subunits to co-ordinate the control of force production. Indeed, analysis of mutations in the β and γ chains, which suppress the paralysis caused by defects in the radial-spoke and central-pair complexes, supports this interpretation. Interestingly, the αβ dimer and the single β and γ heavy chains have different microtubule-binding and -translocation properties [5]. This raises the important question of how these different motor properties can be co-ordinated into a co-operative action leading to a functional power stroke. Other studies have found that the α heavy chain is the only phosphorylated component within the *Chlamydomonas* outer dynein arm. This heavy chain contains at least six phosphorylated sites, most of which are in the head domain. Perhaps importantly, one of these residues is adjacent to the ATP

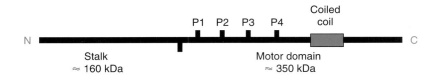

Figure 3. Organization of dynein heavy chains
The diagram shows the key structural motifs common to axonemal and cytoplasmic dynein heavy chains. The N-terminal region consists of an ≈160 kDa stalk domain to which the intermediate chains and other components attach. The large globular motor region contains four nucleotide-binding P-loop consensus sequences (P1–P4) and a short coiled-coil region that appears to mediate microtubule binding.

hydrolysis site while an additional phosphorylated region was identified near to the coiled-coil domain towards the C-terminus (Figure 3). These locations, in concert with the high rates of turnover, suggest that reversible phosphorylation/dephosphorylation has an important regulatory role in dynein function. This hypothesis is supported further by the observation that the rate at which *Paramecium* dynein translocates microtubules in an assay *in vitro* is controlled by the cAMP-dependent phosphorylation of a 29 kDa light-chain component [6].

Recent studies have made progress towards the identification of the microtubule-binding site within the heavy chain. Deletion of a well-conserved region C-terminal to the P4 loop abolished microtubule-binding activity and confirmed the importance of this heavy-chain segment in dynein activity [7,8]. High-resolution structural studies of the head domain of *Dictyostelium* cytoplasmic dynein have visualized seven to eight globular lobes that surround a large central cavity. There is also a stalk domain protruding from the globular domain that is thought to connect the dynein head to its adjacent microtubule in an ATP-sensitive manner and impart lateral force [9,10]. The microtubule-binding C-terminal region downstream of the P4 loop contains conserved sequences predicted to form α-helical coiled-coils. These coil structures may interact in an anti-parallel manner to form the stalk with a small globular (presumably microtubule-binding) unit at the tip. When this region was expressed *in vitro* it was found to co-sediment with microtubules [8]. The N-terminal region (≈160 kDa) of the heavy chains is much more divergent and may impart additional individuality to the heavy chains by determining which accessory components are able to bind and thus modulate cargo attachment and other regulatory activities.

The intermediate chains

Two specific classes of intermediate chain (≈70–120 kDa) are known at present. Outer-arm dynein from both sea urchin and *Chlamydomonas*, as well as mammalian cytoplasmic dynein, contain two copies of related intermediate-chain proteins. These polypeptides have five or six repeated segments in the C-

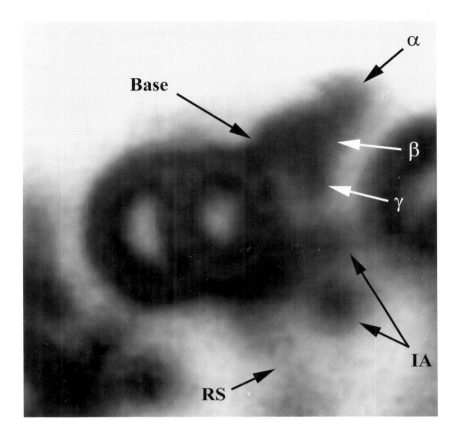

Figure 4. Structure of the outer dynein arm *in situ*
Averaged high-resolution electron micrograph of 16 outer doublet microtubules from
Chlamydomonas flagellar axonemes. The location of the three outer-arm motor domains (α, β
and γ) and the microtubule-binding basal complex are indicated. The inner row of dynein arms
(IA) and the radial spokes (RS) are also indicated. Figure provided kindly by Dr. Ritsu Kamiya
(University of Tokyo, Tokyo, Japan).

terminal region and are members of the WD-repeat protein family (Figure 5).
The WD-repeat motif is ≈40 residues long and the three-dimensional
structure of one protein containing these repeats (the G_β subunit of
heterotrimeric G-proteins) has been solved [11]. Each repeat provides four β
strands that come together to form the blades of a toroidal structure. The N-
termini of each intermediate chain are apparently unrelated and thus far each
appears to exhibit a distinct function. For example, IC1 (formerly IC78 or
IC80) from the *Chlamydomonas* outer arm interacts directly with tubulin *in
situ* [12]. Likewise, IC74 of cytoplasmic dynein binds directly to the
p150*Glued* component of dynactin (an activator of dynein-mediated vesicular
transport) [13]. Electron-microscopic studies of sea urchin and
Chlamydomonas dyneins have revealed that the intermediate chains are located
at the base of the soluble dynein particle. Furthermore, *Chlamydomonas* null

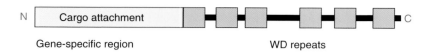

Figure 5. Intermediate-chain organization
All dyneins constructed from two or more heavy chains also contain two intermediate chains that are members of the WD-repeat protein family. The diagram illustrates the location of the key structural motifs of the intermediate chains. The C-terminal portion contains five or six copies of the WD-repeat. Extrapolating from the structure of the G_β subunit of heterotrimeric G-proteins suggests that the WD-repeat is expected to form a toroidal β-propeller structure. The N-terminal region of intermediate chains is gene-specific with little similarity between the various chains. This region is involved in cargo attachment in IC1 from *Chlamydomonas* outer-arm dynein.

mutants defective for either IC1 or IC2 (formerly IC69 or IC70) fail to incorporate outer dynein arms into the axonemal superstructure, indicating that these components are essential for assembly. Together, these observations have led to the hypothesis that the intermediate chains are involved in the attachment of the dynein motor to its specific cargo.

The flagellar dynein intermediate chains are not post-translationally modified. However, IC74 of rat brain cytoplasmic dynein is encoded by two separate genes and is subject to both alternative splicing and differential phosphorylation. Together, these modifications generate multiple isoforms that are developmentally regulated. This suggests that different intermediate-chain isoforms are required at different times during brain development, possibly in concert with changes in cargoes for retrograde axonal transport and/or other dynein-mediated activities [14].

The light chains

The light chains (<30 kDa) are by far the most structurally and functionally diverse dynein components. There are essentially two classes of light chain: those associated with the intermediate chains at the base of the particle (see below) and those that are bound tightly to individual heavy chains. In *Chlamydomonas*, this latter group includes two Ca^{2+}-binding proteins, centrin (associated with inner arms I2 and I3) and LC4 (bound to the outer-arm γ heavy chain). As both phototactic steering (differential regulation of flagellar beat frequency) and the photoshock response (which involves a reversal in swimming direction) are Ca^{2+}-mediated, it is likely that these light chains are involved in the control of specific motor functions. In the *Chlamydomonas* outer arm, both α and β heavy chains interact with members of the thioredoxin superfamily. These polypeptides are redox-active and indeed appear to be functional sulphydryl oxidoreductases. As mentioned above, one *Paramecium* light chain is phosphorylated in a cAMP-dependent manner and appears also to control dynein activity in that organism.

The remaining light chains are located at the base of the dynein particle and are more generic in nature as both cytoplasmic and flagellar dyneins contain similar (and in some cases identical) chains. This group includes the highly conserved LC8 protein and several members of the Tctex1 protein family (these polypeptides are discussed in more detail below). Disruption of the genes encoding some of these proteins leads to the failure of the dynein particle to assemble, emphasizing the importance of these components in dynein structure.

Recent advances

LC8 and retrograde intraflagellar transport

Probably the most fascinating of the dynein light chains is LC8, which was initially identified in *Chlamydomonas* outer-arm dynein [15]. It was subsequently found to be an integral component of mammalian cytoplasmic dynein, inner arm I1 and the unconventional actin motor, myosin V. LC8 has also been identified in yeast two-hybrid screens as physically interacting with a number of additional proteins, including neuronal nitric oxide synthase and $I_\kappa B\alpha$ (inhibitor of nuclear factor κB). Indeed, LC8 has been reported to inhibit nitric oxide synthase activity. Intriguingly, LC8 expression is up-regulated in neuronal cells that are at risk from damage by nitric oxide subsequent to cerebral ischaemia. In *Drosophila*, partial loss-of-function mutants in LC8 result in morphogenetic defects and female sterility, whereas total loss-of-function is embryonic lethal. Together, these observations have led to the suggestion that LC8 plays an important regulatory function in a large number of different systems in a manner analogous to calmodulin.

Intraflagellar transport was originally identified in *Chlamydomonas* as the bidirectional movement of raft-like particles along the flagellar microtubules, in a process independent of flagellar beating [16]. The rafts are located just underneath the flagellar membrane and appear to contain materials involved in the maintenance of flagellar integrity. Movement of rafts away from the cell body is powered by a kinesin-like protein that is encoded at the *fla10* locus. Transport of the rafts back to the cell body is predicted to require a minus-end-directed microtubule motor and it has recently been found that the mutant *fla14* lacks this activity [17]. Flagella from the *fla14* mutant are very short, immotile and have many morphological defects. Most importantly they also have large accumulations of the raft-like particles at the tips. *fla14* actually encodes the *Chlamydomonas* LC8 protein, suggesting that retrograde intraflagellar transport is mediated by a cytoplasmic dynein which is disrupted in the mutant.

Docking complex

Reconstitution experiments have revealed that for the successful assembly of the outer arm on to dynein-depleted axonemes, an additional particle that sediments at 7 S is required [18]. This particle was subsequently termed the outer-arm dynein-docking complex and is composed of three polypeptides of ≈83, 62.5 and 25 kDa. Axonemes from outer-dynein armless mutants that retain the docking complex have small projections at the correct location on the outer doublets, which therefore may represent the docking particle. Structural analysis of docking-complex components identified extensive coiled-coil regions in both the ≈83 and 62.5 kDa polypeptides. The larger protein also contains stretches of highly basic and highly acidic residues that may provide sites for mediating interactions between tubulin of the outer doublets and the basal domain of the outer dynein arm. This flagellar structure appears to be analogous in function to dynactin, which promotes the interaction between cytoplasmic dynein and its vesicular cargo.

Motility *in vitro* and control of dynein function

It has proved possible to assess dynein motor function in two relatively simple assays *in vitro*. One entails coating microscope coverslips with purified dynein, adding ATP and microtubules, and then directly visualizing microtubule translocation across the coverslip [19]. The second method examines dynein function *in situ* by performing controlled proteolysis of flagellar axonemes in the presence of ATP. Proteolysis breaks the linkage holding the axonemal doublet microtubules together and allows them to be moved with respect to each other, thus leading to the sliding disintegration of the entire structure.

The various inner and outer dynein arm species cause microtubules to glide with a large range of velocities [20]. This suggests that different dynein components have different functions in flagellar beating; for example, the inner arms exhibiting the faster gliding velocities may be involved in bend initiation. Also, a subset of the inner arms has been observed to generate torque and thus rotate the gliding microtubules. The physiological significance of this activity is unknown, although it may be involved in the formation of complex flagellar waveforms.

The sliding disintegration assay has yielded important clues as to how dynein arms are regulated *in vivo* by measuring differences in sliding disintegration velocity of axonemes from numerous mutant strains. Dynein inner arms derived from a strain lacking radial spokes translocate microtubules in the sliding disintegration assay at a slower velocity than inner arms from a radial-spoke-containing strain. This indicates that the radial spokes activate the inner arms and that this activated state is maintained even after reconstitution of these active dynein particles into radial-spoke-less axonemes [21].

Dynein and meiotic drive of the murine *t* complex

Recent observations indicate that several dynein light chains are proteins previously cloned as candidates for involvement in meiotic drive exhibited by the murine *t* haplotypes, which is thought to be caused by defects in

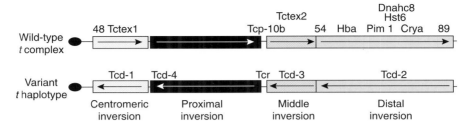

Figure 6. The *t* complex region of mouse chromosome 17
Genomic organization of the two variant forms of the proximal region of chromosome 17: the upper map shows the wild-type *t* complex, while the lower map illustrates the rearrangements present in the variant *t* haplotypes. The large inversions in the *t* haplotype serve to suppress recombination with the wild-type *t* complex. A number of genes and other DNA markers have been located in this region, including the dynein light chains *Tctex1* and *Tctex2*, the hybrid-sterility locus (*Hst6*) and a presumptive axonemal dynein heavy chain (*Dnahc8*). On the *t* haplotype map are the locations of the distorter (*Tcd1–4*) and responder (*Tcr*) loci that are mutated in the *t* haplotype and are responsible for the non-Mendelian transmission of the *t* haplotype-containing copy of chromosome 17. Modified from Pilder et al. [24] with permission. ©1993, Academic Press.

spermiogenesis [22,23]. The *t* complex encompasses the proximal 30–40 Mb of mouse chromosome 17 (Figure 6). In wild mouse populations variant forms of this region exist (known as *t* haplotypes), characterized by four large inversions that prevent the haplotype from undergoing recombination with the wild-type *t* complex. Therefore, the *t* complex (or haplotype) is transmitted in its entirety to the offspring. The fascinating feature of this system is that heterozygous (+/*t*) males pass the *t* haplotype form of chromosome 17 to essentially all of their progeny in a non-Mendelian process known as transmission ratio distortion or meiotic drive. Genetic analysis has implicated mutations in a series of interacting distorter and responder proteins that lead to defective sperm bearing specifically the wild-type genotype. Therefore, *t* haplotype-containing sperm have a major competitive advantage during fertilization. Two distorter candidates have now been identified as dynein light chains. Tctex1 is a component of inner arm I1 from *Chlamydomonas* flagella, outer-arm dynein from sea urchin sperm and rat brain cytoplasmic dynein. The *Chlamydomonas* outer arm also contains a light chain homologous to a second candidate distorter Tctex2. Importantly, cytoplasmic dynein containing Tctex1 is present in many tissues but the phenotype due to the *t* haplotype-encoded mutations is observed only in the male germ cells. This suggests that these mutations specifically affect the activity of Tctex1 and Tctex2 in axonemal dyneins, which could lead directly to flagellar dysfunction. This has been the first indication that alterations in flagellar dynein light chains can result in aberrant sperm motility and consequently profoundly affect male fertility. Clearly much remains to be learnt about this intriguing phenomenon, including the molecular identity of the responder, which is thought to be responsible for differential distribution of wild-type and *t* mutant dyneins during spermiogenesis.

Summary

- *Recent molecular, genetic and functional studies have led to an unparalleled growth in our understanding of dynein and the roles played by the various polypeptides of these massive macromolecular assemblies.*

- *Dyneins are highly complex 1–2MDa complexes that function as molecular motors and move the cargo to which they are attached towards the minus-end of a microtubule.*

- *Dynein motor function is a property of the heavy chains, whereas the intermediate chains are involved in attachment to the appropriate cargo.*

- *In order for useful work to be obtained, motor and cargo-binding activities must be tightly controlled. Current data suggest that this is the role played by certain accessory light-chain proteins.*

- *The LC8 is highly conserved and found in many enzyme systems. This protein is essential in multicellular organisms.*

- *The dynein light chains Tctex1 and Tctex2 have been implicated in the non-Mendelian transmission of variant forms of mouse chromosome 17.*

We thank Dr. Ritsu Kamiya for the providing the electron micrograph shown in Figure 4 and Dr. Mary Porter for sharing unpublished information. Work in our laboratory is supported by grant GM 51293 from the National Institutes of Health and by the Heritage Affiliate of the American Heart Association. A.H. is a Lalor Foundation Fellow.

References

1. Gibbons, I.R., Gibbons, B.H., Mocz, G. & Asai, D.J. (1991) Multiple nucleotide binding sites in the sequence of dynein β heavy chain. *Nature (London)* **352**, 640–643

2. Koonce, M.P., Grissom, P.M. & McIntosh, J.R. (1992) Dynein from *Dictyostelium*: primary structure comparisons between a cytoplasmic motor enzyme and flagellar dynein. *J. Cell Biol.* **119**, 1597–1604

3. Mocz, G. & Gibbons, I.R. (1996) Phase partition analysis of nucleotide binding to axonemal dynein. *Biochemistry* **35**, 9204–9211

4. Sakakibara, H., Takada, S., King, S.M., Witman, G.B. & Kamiya, R. (1993) A *Chlamydomonas* outer arm dynein mutant with a truncated β heavy chain. *J. Cell Biol.* **122**, 653–661

5. Sakakibara, H. & Nakayama, H. (1998) Translocation of microtubules caused by the αβ, β and γ outer arm dynein subparticles of *Chlamydomonas. J. Cell Sci.* **111**, 1155–1164

6. Barkalow, K., Hamasaki, P. & Satir, P. (1994) Regulation of a 22 S dynein by a 29-kDa light chain. *J. Cell Biol.* **126**, 727–735

7. Koonce, M.P. (1997) Identification of a microtubule-binding domain in a cytoplasmic dynein heavy chain. *J. Biol. Chem.* **272**, 19714–19718

8. Gee, M.A., Heuser, J.E. & Vallee, R.B. (1997) An extended microtubule- binding structure within the dynein motor domain. *Nature (London)* **390**, 636–639

9. Goodenough, U.W. & Heuser, J.E. (1984) Structural comparison of purified dynein proteins with *in situ* dynein arms. *J. Mol. Biol.* **180**, 1083–1118

10. Samsó, M., Radermacher, M., Frank, J. & Koonce, M.P. (1998) Structural characterization of a dynein motor domain. *J. Mol. Biol.* **276**, 927–937

11. Sondek, J., Bohm, A., Lambright, D.G., Hamm, H.E. & Sigler, P.B. (1996) Crystal structure of a G protein β/γ dimer at 2.1 Å resolution. *Nature (London)* **379**, 369–374

12. King, S.M., Wilkerson, C.G. & Witman, G.B. (1991) The M_r 78,000 intermediate chain from *Chlamydomonas* outer arm dynein interacts with α tubulin *in situ*. *J. Biol. Chem.* **266**, 8401–8407

13. Karki, S. & Holzbaur, E.L. (1995) Affinity chromatography demonstrates a direct binding between cytoplasmic dynein and the dynactin complex. *J. Biol. Chem.* **270**, 28806–28811

14. Pfister, K.K., Salata, M.W., Dillman, III, J.F., Vaughan, K.T., Vallee, R.B., Torre, E. & Lye, R.J. (1996) Differential expression and phosphorylation of the 74-kDa intermediate chains of cytoplasmic dynein in cultured neurons and glia. *J. Biol. Chem.* **271**, 1687–1694

15. King, S.M. & Patel-King, R.S. (1995) The M_r=8,000 and 11,000 outer arm dynein light chains from *Chlamydomonas* flagella have cytoplasmic homologues. *J. Biol. Chem.* **270**, 11445–11452

16. Kozminski, K.G., Johnson, K.A., Forscher, P. & Rosenbaum, J.L. (1993) A motility in the eukaryotic flagellum unrelated to flagellar beating. *Proc. Natl. Acad. Sci. U.S.A.* **90**, 5519–5523

17. Pazour, G.J., Wilkerson, C.G. & Witman, G.B. (1998) A dynein light chain is essential for the retrograde particle movement of intraflagellar transport (IFT). *J. Cell Biol.* **141**, 979–992

18. Takada, S. & Kamiya, R. (1994) Functional reconstitution of *Chlamydomonas* outer dynein arms from α-β and γ subunits: requirement of a third factor. *J. Cell Biol.* **126**, 737–745

19. Paschal, B.M., King, S.M., Moss, A.G., Collins, C.A., Vallee, R.B. & Witman, G.B. (1987) Isolated flagellar outer arm dynein translocates brain microtubules *in vitro*. *Nature (London)* **330**, 672–674

20. Kagami, O. & Kamiya, R. (1992) Translocation and rotation of microtubules caused by multiple species of *Chlamydomonas* inner-arm dynein. *J. Cell Sci.* **103**, 653–664

21. Smith, E.F. & Sale, W.S. (1992) Regulation of dynein-driven microtubule sliding by the radial spokes in flagella. *Science* **257**, 1557–1559

22. Patel-King, R.S., Benashski, S.E., Harrison, A. & King, S.M. (1997) A *Chlamydomonas* homologue of the putative murine *t* complex distorter Tctex2 is an outer arm dynein light chain. *J. Cell Biol.* **137**, 1081–1090

23. Harrison, A., Olds-Clarke, P. & King, S.M. (1998) Identification of the *t* complex-encoded cytoplasmic dynein light chain Tctex1 in inner arm I1 supports the involvement of flagellar dyneins in meiotic drive. *J. Cell Biol.* **140**, 1137–1147

24. Pilder, S.H., Olds-Clarke, P., Phillips, D.M. & Silver, L.M. (1993) *Hybrid sterility-6*: a mouse *t* complex locus controlling sperm flagella assembly and movement. *Dev. Biol.* **159**, 631–642

8

Microtubule-based transport along axons, dendrites and axonemes

Dawn Signor & Jonathan M. Scholey[1]

Section of Molecular and Cellular Biology, University of California at Davis, One Shields Avenue, Davis, CA 95616, U.S.A.

Introduction

Many eukaryotic cells elaborate highly asymmetric cytoplasmic extensions whose assembly, maintenance and function depend upon intracellular transport systems that are based on microtubules (MTs) and MT-associated motor proteins. Examples include the axons and dendrites that protrude from neuronal cell bodies, and the cilia and flagella that extend from diverse cell types. These structures, which are the focus of this brief review, have diverse biological functions. For example, in the nervous system, dendrites and axons are specialized for signal reception and signal transmission, respectively. In contrast, motile and sensory ciliary axonemes have evolved to function either to move cell surfaces relative to fluid media or to participate in sensory transduction.

The formation, maintenance and function of axons, dendrites and axonemes depends upon mechanisms for localizing key components that are concentrated specifically in these structures relative to cell bodies. For example, signal reception by dendrites and sensory cilia depends upon the action of signal receptors and signal-transduction components that are localized specifi-

[1]*To whom correspondence should be addressed.*

cally to the plasma membrane surrounding these structures, and signal transmission by neurons depends on the delivery of synaptic-vesicle components to the axon termini where they accumulate, awaiting release by regulated secretion. Similarly, structures required for the beating of motile cilia and flagella, such as dynein arms, radial spokes and nexin links are localized specifically to axonemes.

How do key components become specifically localized to cytoplasmic extensions? There is no protein synthesis in axons or axonemes, so proteins that are required in these structures are synthesized in the cytoplasm of the cell body, then actively transported to their site of action. For example, some proteins involved in synaptic transmission are synthesized and packaged into vesicles in the neuronal cell body where the rough endoplasmic reticulum and Golgi apparatus reside, and are delivered by active vesicular transport along the axon to the synapse. Similar transport mechanisms are likely to operate in dendrites, although these neuronal extensions, unlike axons, contain ribosomes and translational machinery, so here the localization of mRNAs encoding key proteins may play an important role as well. In cells that elaborate ciliary or flagellar axonemes, axonemal components must be preassembled in the cell body, then transported as macromolecular protein complexes along the axoneme. The transport of membrane-bound organelles and macromolecular complexes along axons, dendrites and axonemes depends upon vectorial transport driven by MTs and MT-associated motor proteins.

MTs and MT-based motor proteins

MT-based motor proteins are enzymes that hydrolyse ATP and use the energy released to move and transport cargo unidirectionally along a MT track (Figure 1). MT tracks are 25-nm-diameter tubular polymers formed by the lateral aggregation of ≈13 parallel protofilaments, with each structurally polar protofilament consisting of a linear array of α,β tubulin heterodimers arranged with an 8 nm periodicity along the filament axis. Thus MTs have plus- (or fast-growing) ends, and minus- (or slow-growing) ends. Most MTs emanate from MT-organizing centres (MTOCs) with their plus-ends distal and their minus-ends proximal to the MTOC.

The motors that move along these MT tracks are linear, capable of stepping sequentially from one binding site to the next along the linear polymer lattice, with each step being coupled to the hydrolysis of an ATP molecule [1] (see Figure 1). MT-based motor proteins have motor domains that are responsible for ATP hydrolysis and MT binding; these domains are attached via linker segments to cargo-binding domains that bind to the membrane-bound vesicles or macromolecular complexes to be transported along the MT tracks (Figure 1).

The mechanism of action of one MT motor, conventional kinesin (see below), has been investigated in detail and the data obtained suggest that it is a

(a)

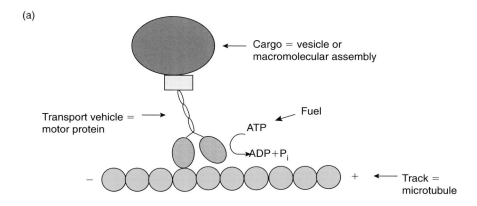

Cargo = vesicle or
macromolecular assembly

Transport vehicle =
motor protein

Fuel

ATP

$ADP+P_i$

Track =
microtubule

(b)

Motor domains

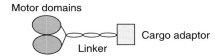

Cargo adaptor

Linker

Figure 1. Design of transport systems
(a) Schematic representation of a typical linear motor protein transporting cargo along an MT protofilament. The translocation of the motor along cellular MTs is an energy-requiring process. (b) Domain structure of a typical linear motor.

processive enzyme that attaches to an MT and moves its vesicular cargo, at ≈1 μm per s, for a distance of ≈1 μm towards the MT plus-end before detaching again. To accomplish this, the motor moves in a hand-over-hand fashion, taking about 100 steps of 8 nm each along the polymer lattice, with each step coupled to the hydrolysis of one ATP molecule and lasting ≈15 ms. Other MT motors also move unidirectionally along MT tracks, carrying their cargo either towards the MT plus-end or towards the MT minus-end. However, it is uncertain if these motors are processive or not.

The organization of MTs in axons, dendrites and axonemes

Our current understanding of the organization of the MT arrays in neurons and ciliated or flagellated cells is depicted in Figure 2. In most neurons, a perinuclear centrosome localized in the cell body functions as the MTOC; MTs emanate from this structure to form a radial cytoplasmic array with the MT plus-ends pointing towards the cell periphery. The organization of MTs in dendrites and axons differ; this may reflect functional specializations of these processes in signal transmission and reception respectively, but how is unclear. For example, in axons MTs are discontinuous and are organized into parallel arrays of uniform polarity, with their minus-ends proximal to the cell body,

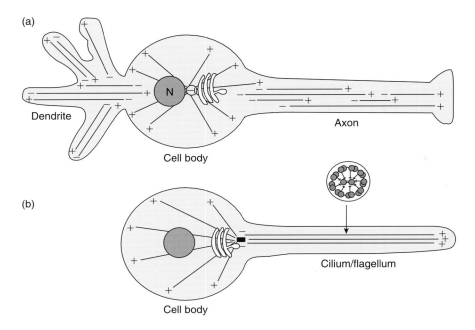

Figure 2. MTs in neuronal processes and axonemes
(a) Representation of a generalized neuron, depicting the three characteristic features of these highly polarized cells: the dendrite (signal-receiving site), the cell body [contains the nucleus (N), MTOC and protein-synthetic/sorting machinery] and the axon (site of synaptic-vesicle secretion and neurotransmitter release). The MTs exist as parallel, discontinuous arrays of uniform polarity in the axon; but, in contrast, are anti-parallel and of mixed polarity in dendrites. (b) Representation of a generic ciliated/flagellated cell. Like neurons, the cell body contains the nucleus and protein-synthetic/sorting machinery; however, both cytoplasmic and axonemal MTs emanate from a basal body located at the base of the cilium/flagellum (shown as a black rectangle). The ciliary/flagellar axoneme comprises nine outer-doublet MTs and a central-pair apparatus, as shown in the cross-section above the cilium/flagellum. The axonemal MTs are of uniform polarity, with their minus-ends proximal to the basal body, and their plus-ends distal.

and their plus-ends distal. In dendrites, however, the MTs are of mixed polarity, being organized into antiparallel arrays. These axonal and dendritic MT arrays are not attached to any detectable organizing structure. In ciliated and flagellated cells, cytoplasmic and axonemal MTs emanate from a basal body, a specialized MTOC that is resorbed during mitosis to serve as the spindle pole. The axonemes of both motile and sensory cilia consist of parallel cylindrical arrays of nine doublet MTs organized with their plus-ends pointing away from the basal body. In motile axonemes, the nine doublet MTs surround two central-pair singlet MTs, and the concerted action of the associated dynein arms, radial spokes and nexin links cause the axomenes to beat in a coherent fashion. Sensory cilia are immotile and have relatively simple structures, lacking dynein and nexin arms, radial spokes and the central-pair apparatus.

The vectorial transport of membrane-bound organelles and macromolecular complexes along the arrays of MT tracks in axons and axonemes is bidirectional: either in anterograde fashion towards MT plus-ends, or in retrograde fashion towards MT minus-ends. Thus given the uniform polarity of MTs in axons and axonemes, there is a requirement for both plus- and minus-end-directed motor proteins. Less is known about transport along the antiparallel arrays of MTs in dendrites, but here again it is thought that plus- and minus-end-directed motors operate. The MT motors that mediate these transport events are members of the kinesin and dynein superfamilies.

MT-based motor proteins of neurons and axonemes

The kinesin and dynein superfamilies constitute a very large group of MT-based molecular motor enzymes that are involved in a wide range of intracellular transport functions (Table 1). Conventional kinesin was the first intracellular transport motor to be identified [2]. Subsequently, molecular techniques have led to the identification of a multitude of related motors. For example, kinesin-related proteins characterized by the presence of the kinesin superfamily motor domain [3] have been identified through the analysis of mutants defective in organelle transport [4], through the use of pankinesin peptide antibodies [5,6] and through the use of the PCR [7,8]. Indeed, PCR has revealed the presence of over 20 distinct kinesin-related polypeptides (KIFS) in the mouse, where they are proposed to participate in various aspects of neuronal-cell function.

Kinesin motor polypeptides are generally composed of three characteristic domains: a motor domain with both MT-binding and nucleotide-hydrolysis functions; a stalk domain that is typically rich in α-helical coiled-coils and facilitates the dimerization of motor polypeptides; and a non-conserved tail domain [3]. The position of the motor domain varies depending on the class of kinesin-related proteins [1]. Those with N-terminal motor domains move predominantly towards the plus-ends of MTs, whereas those with C-terminal motor domains move predominantly towards the minus-ends of MTs. Kinesin motor polypeptides assemble with other motor polypeptides and/or accessory proteins into multimeric motor holoenzymes [9]. For example, conventional kinesin is a heterotetramer comprising two kinesin heavy chains (KHCs) and two kinesin light chains, whereas kinesin-II motor complexes are generally heterotrimeric, containing two different heterodimerized motor subunits and a non-motor accessory polypeptide.

The inventory of kinesin motors involved in intracellular transport has recently been reviewed by Hirokawa [10]. These include: conventional kinesin, a plus-end-directed N-terminal motor responsible for moving exocytotic vesicles and ion channels out to the cell periphery [11,12]; monomeric kinesins that move synaptic vesicles and mitochondria to the plus-ends of MTs [4,10]; heteromeric kinesins, containing at least two distinct N-terminal motor

Table 1. Motors involved in neuronal and axonemal transport

Motor	Oligomeric state	Proposed function
Kinesins		
Conventional kinesin		Fast axonal transport of membrane-bound organelles and vesicles; organelle positioning
Heteromeric kinesins		Multifunctional motor: neuronal and mitotic-vesicle transport; ciliogenesis/IFT; Golgi–endoplasmic reticulum traffic; melanosome transport
Monomeric kinesins		Presynaptic vesicle and mitochondrial transport in neurons
C-terminal kinesins		Retrograde transport of organelles in dendrites
Dyneins		
Cytoplasmic dynein		Retrograde organelle transport and IFT; organelle positioning

IFT, intraflagellar transport.

polypeptides that move towards the plus-ends of MTs and participate in membrane traffic and ciliary transport [6,13,14]; KIF4 kinesins, which are N-terminal homodimeric kinesins that move to the plus-ends of MTs and participate in vesicle transport [10]; I-kinesins, homodimers of two motor subunits containing internal motor domains that participate in anterograde vesicle transport and neurite outgrowth [10]; and C-terminal kinesins (containing a C-terminal motor domain), which move towards the minus-ends of MTs, and are thought to participate in organelle transport along dendrites [10].

The dynein superfamily includes both axonemal and cytoplasmic dyneins, all of which move towards the minus-ends of MTs [10,15]. Axonemal dynein generates the force required to bend motile cilia and flagella, whereas cytoplasmic dynein is involved in several intracellular events including retrograde trafficking in neurons and axonemes. Cytoplasmic dynein is a large-molecular-mass protein complex, comprising two dynein heavy chains with motor activity, and eight to nine non-motor accessory subunits (approximately three intermediate chains, four light intermediate chains, and one or two light chains), that functions in concert with the multimeric dynactin complex.

Organelle transport in neuronal-cell extensions

In neurons, membrane-bound organelles and vesicles must be transported within cell bodies to the base of axons and dendrites, and then moved bidirectionally along the axons and dendrites between the cell bodies and the synapses. Axonal transport includes both the fast axonal transport of membranous vesicles and organelles, and the slow axonal transport of cytoskeletal components. Fast axonal transport occurs in both anterograde and retrograde fashions; whereas slow axonal transport occurs only in an anterograde manner. The anterograde transport of neuronal components is driven by conventional kinesin plus multiple kinesin-related proteins, whereas retrograde transport is driven by dynein and C-terminal kinesins.

Transport of membranous vesicles in neurons

Conventional kinesin is the most abundant motor in neurons. Generally, conventional kinesin is thought to function as a membrane traffic motor in cells, and in neurons it has been implicated in the anterograde transport of axonal vesicles. For example, when mouse peripheral nerves are ligated to block axonal transport, kinesin is found associated primarily with vesicles that accumulate on the side of the ligature proximal to the cell body. Additionally, some mutations in *Drosophila* KHC appear to reduce the number of voltage-gated Na^+ channels in axons [12]. Although the KHC gene is essential in *Drosophila*, certain hypomorphic alleles cause a paralysis phenotype that is amenable to analysis in larvae [16]. In these mutants, axon terminals are reduced in size, axon-potential propagation is impaired and reduced neurotransmitter release is observed at nerve termini; however, no effects on the localization of synaptic vesicles at the axon terminal are seen [16]. Mutations in *Drosophila* kinesin light chain display the same phenotype as that observed in *khc* mutants, suggesting that the associated light chains are essential for axonal transport and kinesin function. Conventional kinesin was initially an attractive candidate motor for the fast anterograde transport of presynaptic vesicles, but these data suggest that it may instead be important for the transport and delivery of exocytotic vesicles containing ion channels to the synaptic terminal, possibly by a two-step mechanism involving kinesin-driven transport along axonal MTs followed by myosin-mediated transport along cortical actin filaments [11].

There is now good evidence that presynaptic vesicles are transported instead by members of the monomeric subfamily of kinesin-related proteins. Hall and Hedgecock [4] demonstrated that the monomeric kinesin, UNC-104, in the nematode *Caenorhabditis elegans* is required for the transport of presynaptic vesicles, but not other membrane-bound organelles in neurons. *unc-104* mutants have few synaptic vesicles in their axon termini, and show an accumulation of presynaptic vesicles in cell bodies [4]. Thus UNC-104 is responsible for the anterograde transport of presynaptic vesicles from the trans-Golgi network to the axonal terminal. More recently, an UNC-104 homologue was identified in mice [10], namely KIF1A. KIF1A is a neuron-specific motor that is enriched in axons, and is associated with vesicles containing synaptic-vesicle proteins such as synaptophysin, synaptotagmin and Rab3. KIF1A is essential, as knock-out mice die shortly after birth. Their neurons have decreased synaptic-vesicle precursors, a reduction in synaptic-vesicle density and a corresponding accumulation of small clear vesicles in the cell body.

The heteromeric kinesins have also been implicated in vesicle transport in neurons. In mice, two distinct forms of heterotrimeric kinesin-II complexes are seen: one that is neuron-specific (KIF3A/KIF3C/KAP3), and one that is expressed in both neuronal and non-neuronal cells (KIF3A/KIF3B/KAP3). Immuno-isolation and subcellular-fractionation experiments suggest that both forms are associated with membranous vesicles and membrane-containing cytosolic fractions in murine neurons, the latter complex being found to be associated with 90–160 nm axonal vesicles. Additionally, a heteromeric kinesin homologue is expressed throughout the central and peripheral nervous systems in *Drosophila*. These data suggest that heteromeric kinesin family members are also involved in the fast anterograde neuronal transport of membranous cargo.

Although cytoplasmic dynein has traditionally been viewed as the minus-end motor for retrograde transport in axons, a C-terminal kinesin (KIFC2) has recently been implicated in the retrograde transport of membranous organelles towards the minus-ends of MTs in the mouse nervous system [10]. Immuno-isolation of KIFC2 and its bound cargo suggest that it transports multivesicular body-like organelles in adult-mouse neurons [10]. It is likely that the retrograde transport activities of KIFC2 and cytoplasmic dynein, and the anterograde transport activities of conventional kinesin and other kinesin-related proteins, drive the movement of distinct populations of organelles, thereby allowing the differential regulation of trafficking within axons and dendrites.

Transport of mitochondria in axons

The transport of mitochondria between the cell body and axon is bidirectional. Recent evidence suggests that KIF1B, an N-terminal monomeric kinesin, is involved in the anterograde transport of mitochondria in mouse neurons [10]. It co-localizes with mitochondria, and co-purifies with mitochondria from cultured neuronal cells, but is not detected in cytosolic fractions enriched in synaptic vesicles. Recombinant KIF1B is capable of transporting purified mitochondria *in vitro*.

Transport in dendrites

Whereas abundant evidence exists for the role of kinesins and dynein in axonal transport, little is known about specific transport processes in dendrites. This is presumably because results are more difficult to interpret given the mixed polarity of their MTs. However, recent work suggests that kinesins may be required for dendritic differentiation in rat sympathetic neurons. CHO1/MKLP1 is a kinesin motor expressed in the dendrites of terminally differentiated rat sympathetic and hippocampal neurons [17]. Overexpression of the motor domain of this kinesin in insect ovarian cells led to the induction of dendrite-like processes with non-uniform MT orientation, whereas antisense perturbation of CHO1/MKLP1 expression in rat sympathetic and hippocampal neurons suppressed dendritic differentiation [17,18]. Thus the transport of preassembled MTs from the cell body of the neuron into dendrites and axons, with either their plus- or minus-ends leading, probably establishes the non-uniform array of dendritic MTs in neurons [17,18].

At least two kinesin motors are implicated in dendritic transport of membrane-bound organelles and vesicles. The C-terminal kinesin, KIFC2, is localized to neuronal-cell bodies and dendrites in a manner consistent with a role in the transport of multivesicular bodies from the cell body along dendritic extensions by moving these organelles towards the minus ends of MT tracks. Additionally, the recently described N-terminal kinesin, KIF21B, is also specifically enriched in dendrites.

Membrane trafficking in the neuronal cell body

Recent evidence suggests that conventional kinesin is involved in the maintenance and structure of the endoplasmic reticulum and Golgi apparatus in neuronal-cell bodies. When KHC expression is inhibited in cultured hippocampal neurons and astrocytes with antisense oligonucleotides, a less extensive network of endoplasmic reticulum and Golgi cisternae is seen, with a corresponding collapse of the Golgi complex and endoplasmic reticulum into the nucleus [19]. Studies in cultured neurons have also suggested a role for conventional kinesin in the endocytic pathway. The normal transport of endosomes from the cell periphery to a final perinuclear position requires intact MTs, and occurs in the retrograde direction. However, late endosomes and lysosomes have the capacity to move bidirectionally. The anterograde movement of lysosomes can be induced experimentally with changes in pH: cytoplasmic alkalination causes the perinuclear clustering of lysosomes, which are dispersed subsequently in an anterograde fashion towards the cell periphery with cytoplasmic acidification. This pH-induced dispersion of lysosomes is inhibited in cultured hippocampal neurons with antisense perturbation of KHC expression [19]. Whereas conventional kinesin may be important in the anterograde movement of lysosomes, the more common retrograde movement of endosomes and lysosomes requires the activity of a minus-end motor. Indeed, cytoplasmic dynein has been localized to endocytic vesicles by immunoelectron microscopy.

Transport along ciliary and flagellar axonemes

Cilia and flagella are specialized MT structures that can be adapted for motility and sensory functions (Figures 2 and 3). Because no protein-synthetic machinery exists within the axoneme, its components must be synthesized in the cytoplasm, then actively transported as preassembled intermediates into the cilium or flagellum to their site of assembly. Various proteins are required in motile cilia and flagella, including MTs, non-tubulin axonemal components such as dynein arms and radial spokes, and axoneme-stabilizing proteins. In immotile sensory cilia, accessory structures such as dynein arms and radial spokes that are required for the beating of motile axonemes are not needed, but the sensory receptors and sensory signal-transduction components that are concentrated in these cilia must be transported into sensory cilia from their site of synthesis in the cell body. Increasing evidence indicates that members of the heteromeric subfamily of kinesins and cytoplasmic dynein are important for the bidirectional transport of axonemal components along axonemal MTs, in a process called intraflagellar transport (IFT) [14].

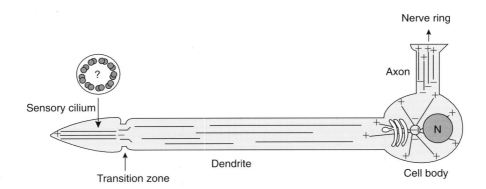

Figure 3. Chemosensory neuron of C. elegans
These neurons have dendrites that terminate in immotile sensory cilia that are open to the external environment through pores in the body cuticle. The polarity and organization of the dendritic and axonal MTs are unknown. The sensory cilia comprise non-motile axonemes that have nine outer-doublet MTs and a variable number of inner-singlet MTs, indicated by '?', which are presumably of uniform polarity with minus-ends proximal to the transition zone (between dendrite and cilium), and plus-ends distal.

IFT

IFT has been studied elegantly in the unicellular biflagellate green alga *Chlamydomonas*. IFT involves the bidirectional transport of electron-dense raft particles beneath the flagellar membrane along outer MT doublets [20]. These raft particles are non-membrane-bound macromolecular complexes that are transported along the entire length of the flagellum by a process dependent on the product of the *fla10* gene. The Fla10 protein is a subunit of a heterotrimeric kinesin-II motor complex [6,21] which immunolocalizes to the basal bodies and proximal portions of the flagella in interphase cells, and in punctate structures that extend to the tip of the developing axoneme during flagellar regeneration [21]. The *fla10* mutant demonstrates temperature-sensitive defects in flagellar synthesis and maintenance, and a reduction in IFT that is concomitant with a disappearance of the raft particles from the shaft of the axoneme [20]. IFT raft particles have been characterized biochemically, and shown to include 15 polypeptides, two of which have known homologues in *C. elegans* (OSM-1 and OSM-6), which are required for sensory ciliogenesis in chemosensory neurons [21]. Thus the heterotrimeric Fla10 kinesin is responsible for the anterograde transport of raft particles comprising components required to build and maintain motile flagella in *Chlamydomonas*.

Retrograde transport of raft particles appears to be dependent on cytoplasmic dynein. The *fla14* gene encodes a dynein light chain, and *fla14* mutants demonstrate bulges on the sides or tips of the flagella. These bulges represent an accumulation of raft particles that have been transported in an anterograde fashion into the flagellum by Fla10, but are incapable of returning due to the absence of retrograde movement [22]. Interestingly, one of the cargo molecules transported on raft particles by Fla10 appears to be inner dynein arms, which are required by axonemal dynein to generate the force required in the formation of flagellar waveforms.

Ciliogenesis

As in flagellar synthesis, ciliogenesis requires the transport of axonemal components from the cytoplasm to their site of assembly. In the developing sea urchin, motile cilia are formed on blastula stage cells, and are used for embryonic swimming and larval feeding. When fertilized urchin eggs were micro-injected with monoclonal antibodies against one of the motor subunits of heterotrimeric kinesin II, a dramatic inhibition of ciliogenesis was observed, resulting in the formation of truncated, paralysed cilia [23]. Detailed analysis of injected embryos suggests that kinesin II is responsible for the transport of ciliary components that are required to elongate and confer motility to a non-motile 'procilium'. This procilium is thought to be a ciliary assembly intermediate that forms by a kinesin-II-independent process [23].

Sequential transport along dendrites and axonemes in *C. elegans* chemosensory neurons

Heteromeric kinesins are also involved in the assembly of sensory cilia. For example, the sensory nervous system of *C. elegans* comprises a subset of neurons with dendritic processes terminating in non-motile cilia that are open to the external environment (Figure 3). These sensory cilia serve as specialized compartments to localize sensory receptors and signal-transduction machinery, which recognize specific chemical cues in the environment. Therefore, *C. elegans* chemosensory neurons are interesting examples of polarized cells that couple dendritic transport, ciliogenesis and ciliary transport.

Heteromeric kinesins are candidate motor proteins for transporting sensory ciliary components from their site of synthesis and assembly in the cell body out to the dendritic endings. Recently, two distinct heteromeric kinesin complexes have been found in chemosensory neurons and sensory cilia in *C. elegans*, namely heterotrimeric kinesin II and dimeric OSM-3 kinesin [24]. In addition, *C. elegans* mutants have been isolated that display defects in chemosensory behaviour, many of which show ultrastructural defects in the axonemal structures that comprise the sensory cilia [25,26]. One of these mutants, *osm-3*, displays defects in osmotic avoidance and chemotaxis behaviour, truncated sensory cilia and an accumulation of electron-dense vesicles in the surrounding sheath cells [25]. The *osm-3* gene encodes the motor subunit of the dimeric OSM-3 kinesin [24]. Thus OSM-3 kinesin is required for the transport and delivery of cargo, possibly axonemal intermediates, that are required to build or maintain a sensory cilium in *C. elegans* dendrites. Two other mutants, *osm-1* and *osm-6*, also display chemosensory and ciliary defects. These genes encode homologues of proteins identified in IFT raft particles in *Chlamydomonas*, which are transported by the heterotrimeric Fla10 kinesin-II motor protein [21,25]. Recent work using a transport assay *in vivo* has demonstrated that fluorescently labelled OSM-1 and OSM-6 proteins move at the same velocity as labelled kinesin II in sensory cilia, suggesting that they are likely cargoes of the *C. elegans* heterotrimeric kinesin-II complex [27]. Therefore, the two heteromeric kinesins, kinesin II and OSM-3 kinesin, are probably involved in complementary and/or overlapping transport pathways in *C. elegans* IFT [24].

Summary

- *MTs in cytoplasmic extensions including axons, dendrites and axonemes serve as polarized tracks for vectorial intracellular transport driven by MT-based motor proteins.*

- *Although axons and axonemes serve very different functions, increasing evidence suggests that the transport events, MT organization and the motors involved in their formation and function are conserved. Thus, there are obvious similarities in the mechanisms of axonal transport and IFT.*

- *The MT arrays of axons and axonemes are parallel, whereas those of dendrites are anti-parallel, but the functional significance of this difference and its consequences for mechanisms of transport along these processes are unclear.*

- *MT-based motor proteins of the dynein and kinesin superfamilies transport a variety of cargos including membrane-bound vesicles and macromolecular complexes along MTs of axons, dendrites and axonemes, and thus contribute to the formation, maintenance and function of these cytoplasmic extensions.*

- *Chemosensory neurons in the nematode C. elegans represent an appealing system for studying transport events along dendrites and axonemes that occur sequentially in a single cell.*

References

1. Vale, R.D. & Fletterick, R. (1997) The design plan of kinesin motors. *Ann. Rev. Cell Dev. Biol.* **13**, 745–777

2. Vale, R.D., Reese, T.S. & Sheetz, M.P. (1985) Identification of a novel force-generating protein, kinesin, involved in MT-based motility. *Cell* **42**, 39–50

3. Yang, J.T., Laymon, R.A. & Goldstein, L.S. (1989) A three-domain structure of kinesin heavy chain revealed by DNA sequence and microtubule binding analysis. *Cell* **56**, 879–889

4. Hall, D.H. & Hedgecock, E.M. (1991) Kinesin-related gene unc-104 is required for axonal transport of synaptic vesicles in *C. elegans. Cell* **65**, 837–847

5. Cole, D.G., Cande, W.Z., Baskin, R.J., Skoufias, D.A., Hogan, C.J. & Scholey, J.M. (1992) Isolation of a sea urchin egg kinesin-related protein using peptide antibodies. *J. Cell Sci.* **101**, 291–301

6. Cole, D.G., Chinn, S.W., Wedaman, K.P., Hall, K., Vuong, T. & Scholey, J.M. (1993) Novel heterotrimeric kinesin-related protein purified from sea urchin eggs. *Nature (London)* **366**, 268–270

7. Stewart, R.J., Pesavento, P.A., Woerpel, D.N. & Goldstein, L.S.B. (1991) Identification and partial characterization of six members of the kinesin superfamily in *Drosophila. Proc. Natl. Acad. Sci. U.S.A.* **88**, 8470–8474

8. Aizawa, H., Sekine, Y., Takemura, R., Zhang, Z., Nangaku, M. & Hirokawa, N. (1992) Kinesin family in murine central nervous system. *J. Cell Biol.* **119**, 1287–1296

9. Cole, D.G. & Scholey, J.M. (1995) Structural variations among the kinesins. *Trends Cell Biol.* **5**, 259–262

10. Hirokawa, N. (1998) Kinesin and dynein superfamily proteins and the mechanism of organelle transport. *Science* **279**, 519–526

11. Bi, G.Q., Morris, R.L., Lioa, G., Alderton, J.M., Scholey, J.M. & Steinhardt, R.A. (1997) Kinesin and myosin-driven steps of vesicle recruitment for Ca-regulated exocytosis. *J. Cell Biol.* **138**, 999–1009

12. Hurd, D.D., Stern, M. & Saxton, W.M. (1996) Mutation of the axonal transport motor kinesin enhances *paralytic* and suppresses *shaker* in *Drosophila. Genetics* **142**, 195–204

13. Scholey, J.M. (1996) Kinesin-II, a membrane traffic motor in axons, axonemes, and spindles. *J. Cell Biol.* **133**, 1–4

14. Rosenbaum, J.L., Cole, D.G. & Diener, D. (1999) Intraflagellar transport: the eyes have it. *J. Cell Biol.* **144**, 385–388

15. Holzbauer, E. & Vallee, R.B. (1994) Dyneins: molecular structure and cellular functions. *Ann. Rev. Cell Biol.* **10**, 339–372

16. Guo, M., McDonald, K., Ganetsky, B. & Saxton, W.M. (1992) Effects of kinesin mutations on neuronal functions. *Science* **258**, 313–316

17. Sharp, D.J., Kuriyama, R. & Bass, P.W. (1996) Expression of a kinesin-related motor protein induces Sf9 cells to form dendrite-like processes with nonuniform MT polarity orientation. *J. Neurosci.* **16**, 4370–4375

18. Sharp, D.J., Yu, W., Ferhat, L., Kuriyama, R., Rueger, D.C. & Bass, P.W. (1997) Identification of a MT-associated motor protein essential for dendritic differentiation. *J. Cell Biol.* **138**, 833–843

19. Feiguin, F., Ferreira, A., Kosik, K.S. & Caceres, A. (1994) Kinesin-mediated organelle translocation revealed by specific cellular manipulations. *J. Cell Biol.* **127**, 1021–1039

20. Kozminski, K.G., Beech, P.L. & Rosenbaum, J.L. (1995) The *Chlamydomonas* kinesin-like protein FLA10 is involved in motility associated with the flagellar membrane. *J. Cell Biol.* **131**, 1517–1527

21. Cole, D.G., Diener, D.R., Himelblau, A.L., Beech, P.L., Fuster, J.C. & Rosenbaum, J.L. (1998) *Chlamydomonas* kinesin-II-dependent intraflagellar transport (IFT): IFT particles contain proteins required for ciliary assembly in *Caenorhabditis elegans* sensory neurons. *J. Cell Biol.* **141**, 993–1008

22. Pazour, G., Wilkerson, C.G. & Witman, G.B. (1998) A dynein light chain is essential for the retrograde particle movement of intraflagellar transport (IFT). *J. Cell Biol.* **141**, 979–992

23. Morris, R.L. & Scholey, J.M. (1997) Heterotrimeric kinesin-II, is required for the assembly of motile 9+2 ciliary axonemes in sea urchin embryos. *J. Cell Biol.* **138**, 1009–1022

24. Signor, D., Wedaman, K.P., Rose, L.S. & Scholey, J.M. (1999) Two heteromeric kinesin complexes in chemosensory neurons and sensory cilia of *Caenorhabditis elegans*. *Mol. Biol. Cell* **10**, 345–360

25. Perkins, L.A., Hedgecock, E.M., Thomson, J.N. & Culotti, J.G. (1986) Mutant sensory cilia in the nematode *Caenorhabditis elegans*. *Dev. Biol.* **117**, 456–487

26. Starich, T.A., Herman, R.K., Kari, C.K., Yeh, W.H., Schackwitz, W.S., Schulyer, M., Collet, J., Thomas, J. & Riddle, D. (1995) Mutations affecting chemosensory neurons of *Caenorhabditis elegans*. *Genetics* **139**, 171–188

27. Orozco, J.T., Wedaman, K.P., Signor, D., Rose, L.S. & Scholey, J.M. (1999) Movement of motor and cargo along cilia. *Nature (London)* **398**, 674

9

The molecular motors of cilia and eukaryotic flagella

David Woolley

Department of Physiology, School of Medical Sciences, University of Bristol, Bristol BS8 1TD, U.K.

Introduction

The motile cellular appendages of eukaryotes are known either as cilia or flagella, according to their length and pattern of movement. The distinction need not be emphasized, for the mechanism of force generation is essentially the same: both contain the same internal 'engine', an axoneme about 0.2 μm in diameter. In typical cilia and flagella, the archetypal '9+2' axoneme consists of a cytoskeletal cylinder of microtubules and related proteins linked together dynamically by a class of molecular motors known as the axonemal dyneins. The 9+2 axoneme is precisely assembled and has persisted with very little change through evolution. Yet this uniform structure, depending on the same types of motor protein in the same standard layout, can generate a variety of movement patterns in different cells and in changing circumstances. In this chapter, the characteristics of the axonemal dyneins will be explained. Then I shall discuss their operation and how they may be regulated so that co-ordinated function emerges. The architecture of the 9+2 axoneme may best be appreciated from electron micrographs of transverse sections. In Figure 1(a), the axonemal skeleton is shown, without its motor proteins. In Figure 1(b), the axonemal dyneins are included, placed as arrays of outer and inner arms on each of the nine doublet microtubules. In addition Figure 1 leads to an extremely important conclusion. The complete axoneme (Figure 1b) was taken from a normal human spermatozoon capable of vigorous motility, whereas the axonemes without dynein arms (Figure 1a) were from a man whose spermatozoa were alive but completely immotile (Kartagener's syndrome, a

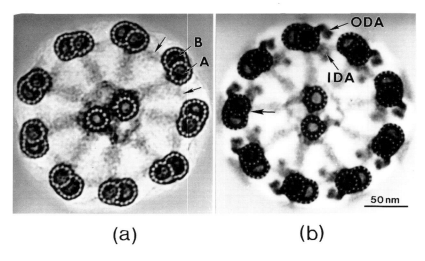

(a)   (b)

Figure 1. Computer-averaged electron micrographs showing transverse sections of axonemes from human spermatozoa

(a) Shows only the cytoskeletal components of the axoneme. There are nine doublet microtubles around the circumference. Each doublet consists of an A-tubule (A, complete) and a B-tubule (B, incomplete). The faint densities between the doublets (arrows) represent the nexin (circumferential) links. From each A-tubule a radial spoke extends towards the central pair of microtubules. (b) The components of the complete, normal axoneme, where dynein arms have assembled into position on the A-tubules. Of the two arrays of dynein arms, the outer arms (ODA) are the more distinct because their periodicity (24 nm) is less than that of the inner arms (IDA, 96 nm). The axoneme here is viewed as if from the base (i.e. from the neck of the spermatozoon.) By convention the doublet having its radial spoke bisecting the central pair (arrow) is number 1; the others are numbered clockwise as viewed from the base. Reproduced from [1], with permission. ©1995, Churchill Livingstone.

type of primary ciliary dyskinesia). Thus the crucial importance of dynein arms for axonemal motility is verified [1]. However, the formal identification of axonemal dynein, the demonstration of its ATPase activity, and the proof that motility depends on it, had been achieved earlier in a series of experiments on the cilia of *Tetrahymena*, begun by Gibbons in 1963 [2]. The position of the dyneins, in the gaps between the doublets, itself suggested a mode of action that is now accepted: that is, a dynein arm, anchored firmly to the A-tubule, is able to exert a force on the B-tubule of the adjacent doublet. This has been established most persuasively through studying the sliding displacements of the adjacent doublet in various situations, both *in vivo* and *in vitro*.

At which points, then, is current knowledge of the function of axonemal motors inadequate? First there is the problem of how any dynein arm exerts force against the opposing tubulin lattice during a cycle of ATP hydrolysis; this is the chemo-mechanical transduction problem common to all of the motor enzymes. The second problem may be posed after noting that each micrometre of axoneme contains over 600 dynein arms of various kinds. What are the temporal and spatial patterns of force generation, both around the circumference and along the length of the axoneme cylinder, that move the organelle in a co-ordinated and repetitive manner? While it is helpful to consider these as separate problems, the force-generating activity of a single dynein arm is unlikely to be an independent event. Mechanical interactions may be expected, perhaps occurring directly between successive dynein arms or indirectly through deformation of the tubulin lattice.

The outer dynein arms (ODAs)

An axoneme is equipped with only one type of ODA, a complex of subunits. There are the catalytic heavy chains, each of ≈500 kDa (two or three according to species, the α-, β- and γ-DHCs, or dynein heavy chains); there are two or three intermediate chains (ICs); and there are six to eight light chains (LCs) [3]. DHC sequences from several organisms have in common a central catalytic domain with four copies of the P-loop consensus motif for nucleotide binding (Figure 2). The P1 loop is the site of cleavage induced by ultraviolet light in the presence of vanadate (a phosphate analogue) and ATP. The sequence of the P1 loop is much more stringently conserved than are those of P2–P4. Mainly for these two reasons, P1 is considered to be the probable location of the ATP hydrolysis that underlies chemo-mechanical transduction [4]. Study of deletion/truncation mutants has located the microtubule-binding domain within the C-terminal end of the DHC, about 180 amino acids downstream of P4, where two regions of coiled-coil α-helix are predicted to align in anti-parallel fashion to form a stalk-like projection [5]. Less is known of the ICs. IC78 from *Chlamydomonas* binds to microtubules *in vitro* and may therefore anchor the arm to the A-tubule [6]. Little is known of the ODA LCs except that LCs from *Chlamydomonas* may also function in association with cytoplasmic dynein [7].

The morphology of the ODA has been highly conserved throughout evolution. After extraction from the axoneme, images of an individual ODA show two or three globular heads, each with a narrow projected stalk; the heads are linked together into a bouquet-like arrangement by a flexible stem that ends in a foot region. The morphology of an ODA *in situ* is considerably more compact. Freeze-etch replicas show a single globular head with usually a single projecting stalk attaching to the B-tubule of the adjacent doublet: whether this represents a conjunction of the two or three heads or a concealment of all but one is not settled. The remainder of the ODA comprises two domains attached to the A-tubule, the proximal (P) foot and the distal (D) foot. There is general consensus that each globular head contains a DHC and it is now very probable that each stalk (or B-link) is a real structure through which the ODA binds to the lattice of the B-tubule [5]. How exactly the P- and D-feet relate to the ICs and LCs is not clear. The linear packing of the ODAs is such that they are spaced 24 nm apart and this requires some overlapping of these domains. In the ATP-depleted state (i.e. in rigor) the head of an ODA obscures its D-foot, whereas, in the relaxed state induced by ATP and vanadate, the head of an

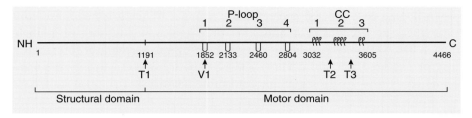

Figure 2. A linear map of the β-heavy chain of sea urchin ODAs
The positions of the P-loops, the site of vanadate-mediated photocleavage (V1), sites of predicted α-helical coiled-coils (CC) and sites of rapid tryptic cleavage (T1–T3) are shown. Reproduced from [4], by courtesy of Marcell Dekker Inc.

ODA obscures the P-foot belonging to the adjacent (proximal) ODA. The conformational change from rigor to relaxed is thus a 12 nm shift of the head in the proximal direction, interpreted as being due to rotation of the head [8]. The entire ODA is therefore seen only where a relaxed ODA has a rigor ODA immediately distal to it (Figure 3).

Two approaches differentiate the functional contributions of the α- and β-DHCs in the overall work of the ODA. From studies of dynein mutations in *Chlamydomonas*, the β-DHC appears to be the more important. Mutations that truncate the β-DHC (*oda-4-s7*; *sup-pf-1*) reduce axonemal beat frequency to a level only slightly greater than the ≈80% reduction found in mutants lacking all parts of the ODA. On the other hand, a mutation (*oda-11*) that selectively disrupts the entire α-DHC gives axonemes that can beat at 85% of

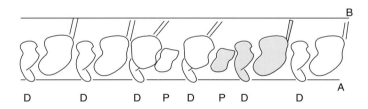

Figure 3. Diagram showing the relationship of ODAs to each other and to the A- and B-tubules, based on averaged electron micrographs of deep-etched replicas of cryo-fixed axonemes
One row of ODAs is shown from the lateral perspective; the ODAs bridge the interdoublet gap between the A-tubule (A) and the adjacent B-tubule (B) at 24 nm intervals. The proximal or basal end of the assembly is on the right, the distal end on the left. Six ODAs are shown, the heads of the dynein arms are the largest features, each having a stalk connecting it to the B-tubule. The P- and D-feet are labelled P and D, respectively. Two rigor ODAs (ATP-depleted state) are shown in the centre of the array, flanked on either side by two relaxed ODAs (believed to represent the step prior to product release and force generation). The shift from rigor to relaxed involves the head making a rotational shift of 12 nm in the proximal direction in preparation for the following power stroke that would drive the B-tubule in the distal direction (i.e. to the left) and restore the rigor arrangement. The ODA is therefore seen in its entirety (light blue) only when it is relaxed and has a rigor arm distal to it. Modified from [8], with permission. © 1995, Academic Press.

the wild-type frequency [3]. The second approach has been to observe the translocation of microtubules *in vitro* over enzyme substrates comprising various fractions of the ODA. The entire ODA supports such gliding motion, as do the β-DHC/IC1 and γ-DHC fractions, but no gliding occurs over a substrate of α-DHC alone [9]. Where translocation *in vitro* has been seen, its direction defines dynein as a minus-end-directed motor [10].

The inner dynein arms (IDAs)

Cryo-electron microscopy shows that axonemes are equipped with three types of IDA arranged in a standard sequence that repeats every 96 nm [11]. The proximo-distal grouping of IDA1, IDA2 and IDA3 lies systematically out of register from one doublet to the next. Like the ODAs, each IDA is a complex of subunits. Eight distinct catalytic heavy chains have been identified and found to be organized with various shorter polypeptides into seven complexes (subspecies a–g), six with a single DHC and one with two DHCs [12]. Some of the DHCs can now be ascribed to specific IDAs and there is currently rapid progress in characterizing the function and position of the various ICs and LCs of the IDAs, the principal method being the production and analysis of motility mutants in *Chlamydomonas*. For example, a mutation in the gene encoding IC140 leads to the failure of assembly of part of IDA1. One of the LCs (LC19) that associates with three of the DHCs is the calcium-binding protein, centrin. Actin is one of the associated LCs in six of the subspecies.

The complex morphology of the IDAs has been conserved to a remarkable extent. Extracted IDAs show, as expected, an unfolded stem–globular head–stalk configuration, some arms having one head, others two [11]. *In situ*, freeze-etch replicas reveal that IDA1 has three heads, IDA2 has two and IDA3 usually has two [11,13]. Figure 4 illustrates the IDAs in the 9+0 sperm flagellum of *Anguilla*. Attempts to demonstrate meaningful structural variations between the rigor and relaxed nucleotide states have not been successful.

As with the ODAs, the functional roles of the various IDAs have been studied in *Chlamydomonas* IDA-assembly mutants and in translocation assays *in vitro*. ODAs are not necessary for co-ordinated flagellar motility; without ODAs the effect is limited to an ≈80% reduction in beat frequency. Therefore, IDA-assembly mutants have been studied in strains also bearing *oda1*, one of 10 mutations that gives a phenotype lacking the entire ODA. On this background it was found that *ida1* and *ida4* caused complete immotility, representing failed assembly of inner subspecies *f* and *acd*, respectively [14]. The presence of ODAs gives motility to *ida1* and *ida4* but not to the double mutant *ida1/ida4*, illustrating that motility is lost when multiple IDA subspecies are missing. All seven subspecies of IDA have been tested for their potential to generate microtubule translocation *in vitro* [12]. Six of them did so, at characteristic velocities, and five subspecies caused rotation of the microtubules as well as forward translation. When IDAs generate microtubule glid-

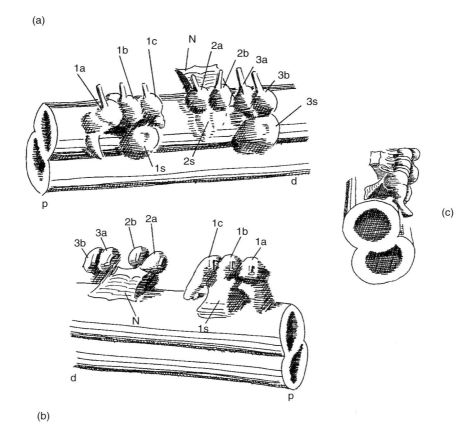

Figure 4. Interpretation of the morphology of the IDA complex in *Anguilla* (eel)
The diagrams, based on freeze-etch replicas, ultrathin sections and whole mounts, give the internal view (a), the external view (b) and a slightly tilted transverse view (c), with the IDAs in blue. The globular heads are labelled 1a–1c (IDA1), 2a and 2b (IDA2), and 3a and 3b (IDA3). In IDA1 the heads protrude from an arched structure supported by a pillar structure [1s in (b)]. This axoneme has naturally neither ODAs nor radial spokes, although attachment sites for the missing spokes can be specified (1s–3s). Proximal (p) and distal (d) orientations are given for (a) and (b) [(c) is drawn from the proximal perspective]. The nexin or circumferential interdoublet linkage is labelled N. The complex of structures shown occupies about 92 nm of the A-tubule and recurs at exactly 96 nm intervals. Reproduced from [13], with permission. © 1997, The Company of Biologists.

ing or power the telescopic sliding of doublets from proteolysed axonemes, the velocities attained are significantly lower than when such movements are powered by ODAs. It seems that the axoneme accommodates dynein arms that have different intrinsic velocities.

Other protein complexes attached to the A-tubule

Three complexes will be considered: the nexin (circumferential) links, the radial spokes (RSs) and the dynein regulatory complex (DRC). There are thought to be functional interrelationships among these three complexes and between them and the dynein motors. They are all related structurally to the IDAs and, like them, they all have a periodicity of 96 nm.

The nexin link bridges the interdoublet gap from the A-tubule to the B-tubule. It lies between the two rows of dynein arms (Figure 1a) close to IDA2 (Figure 4) but it does not resemble a dynein arm. From its connections, it has been assumed to maintain the integrity of the axonemal cylinder and to provide a resistance to interdoublet sliding. If so, the resistance must be frictional rather than elastic because predicted changes in length or angulation were not seen in nexin links when they were examined by the rapid-freeze, deep-etch technique [15].

There are three types of RS (S1, S2 and S3) attached along each of the doublets in most axonemes. The way in which they project on to the central pair (CP) of microtubules is seen in Figure 1. Their attachment sites on the A-tubule may be judged from Figure 4. S1, S2 and S3 attach at the locations labelled 1s, 2s and 3s, respectively. Is the RS–CP abutment essential for bend initiation and propagation? No, because there are instances of these functions being executed in the absence of either (or both) CP and RSs [16]. A possible role for the RSs in regulating the dynein arms will be considered in the next paragraph. It is reasonable to conclude at this point, however, that the RSs must have at least a structural role in preventing the buckling of the axoneme under bending forces.

The DRC is a group of seven polypeptides, several of which are clustered on the A-tubule near IDA2. This suggests a degree of correspondence with the nexin link [13]. Recognition of the DRC followed an investigation of the regulatory role of the RS–CP complex in the 9+2 flagellar axoneme of *Chlamydomonas*, where there are only two types of spoke, S1 and S2. In axonemes lacking ODAs and proteolysed to allow sliding/extrusion of the doublets, the IDAs achieve faster sliding when the RSs are present rather than absent. In non-proteolysed axonemes the absence of the RSs (mutant *pf-14*) or the CP (mutants *pf-15A* or *pf-19*) or the entire RS–CP complex (double mutant) causes paralysis. The paralysis is overcome by any of a series of second-site (suppressor) mutations that affect parts of the ODA, the IDAs or other structures near the IDAs. The current interpretation is that the RSs and CP together generate a signal that relieves an inhibition of dynein activity; the inhibition being due to members of the DRC, as the gene products of the suppressor genes. Thus mutations in the suppressor genes lead to the failure of assembly of the inhibitory components. The biochemistry of the facilitation of dynein by the RS could be that the spokes raise the threshold for substrate inhibition, as spoke-less mutants can swim at lowered ATP concentrations.

The sliding-doublet theory

What happens to the doublet microtubules when the dynein arms generate forces between them? The sliding-doublet theory asserts, most importantly, that such forces cause the doublets to slide linearly rather than to shorten, lengthen or rotate. This theory has guided research on cilia and flagella for 30 years. The essential assertion of the theory is supported now by six lines of evidence: (i) as already documented, purified dynein arms and most of the DHCs cause microtubules to translocate (glide linearly) *in vitro*; (ii) internal measures of doublet length, such as the RS periodicity, show constancy between the convex and concave edges of a bend; (iii) when ATP is added to mildly proteolysed axonemes, the doublets are either extruded telescopically or walk against each other; (iv) linear displacements of predicted magnitude are seen, in electron micrographs, between the doublets at the tips of certain cilia; (v) when localized reactivation is performed by applying ATP iontophoretically, two opposing bends form on either side of the activated region (rather than a single bend as might be expected according to a contraction theory); and (vi) interdoublet displacements consistent with the theory have been detected in reactivated axonemes by measuring the relative movement of attached microbeads [17].

An assembly of axonemal doublets executing strictly rectilinear (one-dimensional) sliding does not, by definition, bend. The theory has therefore had to be extended to explain bend formation, even if the explanations are still speculative. Work on the simplified 9+0 axoneme in the sperm of *Anguilla* [13,16] suggests that an assembly of doublets, IDAs, nexin links and a basal body contains sufficient structures for generating bending moments; but other components may be involved in more complex axonemes.

The theory has had to distinguish active sliding from passive sliding since it became accepted that all the axonemal dyneins operate as unidirectional (minus-end-directed) motors, both *in vitro* [9,10] and where doublets extrude from proteolysed axonemes [3]. Consider an imaginary cylindrical axoneme, with its fixed-length doublets firmly attached at one end to its basal body. Let it become bent into a plane arc without twisting about its own axis. Two-dimensional sliding will have occurred generally between the doublets but the direction (polarity) of sliding will not have been the same between them all. It follows that, in the living axoneme, because of the unidirectionality of the motors, the sliding can have been powered actively only between about half the doublets. Now imagine bending this axoneme in the opposite direction. The opposite set of the doublets will now slide in the direction that corresponds to dynein being active. Thus the dyneins must switch between active and passive status with each cycle of bending [18].

Geometrical deductions from the theory have reached their ultimate elaboration in the analysis of flagella executing planar undulations. Making the simple assumptions indicated in the preceding paragraph, it is possible to calculate, from the waveform and its kinetic parameters, the magnitude and direction of sliding between all nine doublets and their neighbours at any position along the wave, and also the rate of change of sliding displacement with respect to curvature as the bends develop and then travel [19]. The conclusions suggest that the mechanism for controlling the dynein motors must be stupendously complex, with both circumferential and longitudinal regulation imposed reliably and repeatedly through numerous cycles of beating. However, this type of formal analysis has provided the theoretical predictions against which experimental results may be tested, e.g. the experiments using attached microbeads as positional markers [17]. One readily understandable conclusion is that the propagation of a bend is based upon a 'packet' of sliding that propagates with the bend (metachronous sliding). This can be correlated with the observation that zones of tension development propagate spontaneously along groups of doublets that are bathed in ATP. That is, the dynein arms in any position were seen to be recruited into tension development only as the wave of activity reached them [20]. But these geometrical analyses [19] detect other classes of sliding, one of which is difficult to explain: when there is asymmetry between the alternating bends of the planar waveform, as is often the case, it is necessary to conclude that bend growth is accompanied by synchronous sliding throughout the distal flagellum, even though the distal flagellum contains propagating bends. This is not compatible with the common assumption that the sliding–bending transformation is achieved through the localized operation of elastic resistances in which tension develops as active shearing occurs.

The crossbridge theory

This theory states that the dynein arms, anchored firmly to the A-tubule, undergo a conformational change during the hydrolysis cycle to reach across to the opposing B-tubule and interact transiently with it.

Studies on *Tetrahymena* ODAs and singlet microtubles *in vitro*, under conditions in which A-end binding is suppressed, showed that each dynein arm (in fact, each DHC) binds tightly (at its B-end) to the singlet microtubule in the absence of ATP, the rigor state. The binding of ATP rapidly releases this association, with hydrolysis then taking place on the free enzyme. The release of ADP, shown to be the rate-limiting step, was accelerated by reassociation with the microtubule and is thought probably to be the step associated with active sliding. The cycle would then restart with the binding of another molecule of ATP [21].

It was expected that the rigor step would involve a close approximation between the dynein arm and the B-tubule; and that in the ATP-bound and the ADP-bound states the dynein arm would have an extended configuration prior to its power stroke. Electron micrographs of thin sections and whole mounts, respectively, supported these two expectations [22]. However, freeze-etch images have shown that the conformational change is a 12 nm shift, in the proximal direction, of the head domain from the rigor to relaxed state, probably brought about by a rotation of the head domain. Furthermore, the rigor state, which has now been identified in motile axonemes [18], does not give a closer association with the B-tubule.

One power stoke delivered by a transiently attached crossbridge cannot cause the opposing doublet to slide more than ≈16 nm. This follows from the size of the arms. However, the theory of sliding doublets predicts that much greater displacements (≈100 nm) must occur in the formation of a bend, at least between some of the doublet pairs. So some of the dyneins must perform multiple cycles of crossbridge attachment and detachment, and presumably multiple cycles of hydrolysis. One implication is that a detached state should in principle occur. This has been sought among ODAs in axonemes frozen in the motile state but without success. Another implication concerns the ATP-utilization rate per dynein arm per cycle of beating. Experiments have suggested that one or two molecules of ATP are hydrolysed per DHC per beat. This is not high enough to support the idea that multiple hydrolytic steps underlie each sliding displacement and it remains to be discovered how much active sliding is produced by one hydrolysis cycle.

Assessment and prospect

The crossbridge theory gives a plausible molecular basis for the essentially geometrical theory of sliding doublets so that, taken together, these two theories give a coherent and convincing explanation of how the dynein motors function in the motility of cilia and eukaryotic flagella. Perhaps to some extent the theories have found acceptance on account of the very close parallels with muscle motility, but there is no doubt as to the strength of experimental support.

However, the prevailing theories have never explained one of the cardinal features of axonemal motility; its oscillatory quality. In other respects, the theory seems to me to have been developing very slowly in several important areas. If the B-link is the crossbridge, it should act mechanically like a crank arm, although from its morphology it looks more suited to developing tension rather than to resisting bending. The search for the mechanism of the switch, the mechanism that is believed to limit activation to half the dyneins at any location, is still at an early stage; in any case, the scope of this research seems to be restricted to planar-beating axomemes. In addition it is not known how synchronous sliding can be superimposed on metachronous sliding, as the sliding-doublet theory requires, if bending is the result of restrictions on sliding.

There are several recent pieces of experimental work that have not been assimilated into the general theory. Shingyoji et al. [23] used optical trapping manometry to measure the peak force (≈6 pN) exerted by a very few dynein arms (or possibly only one). In addition, they recorded an oscillatory component of the force (≈2 pN peak to peak) with a maximum frequency of ≈70 Hz in 0.75 mM ATP. They suggest that this feature of the individual dyneins, both outer and inner groups, may underlie rhythmic beating.

Another piece of work, a study of sliding disintegration of axonemes, has provided evidence that axonemal dyneins can operate bidirectionally rather than only unidirectionally [24]. The change in direction of active sliding was a result of raising the Ca^{2+} concentration and the reversal of active sliding was believed to alter the chirality of the three-dimensional component of the flagellar wave executed by sea urchin spermatozoa.

The widespread occurrence of three-dimensional beating patterns in cilia and flagella is fully acknowledged yet has rarely been analysed in terms of the internal mechanics. The sliding-doublet theory may apply in principle if one imagines a control system that recruits dyneins into activity circumferentially as well as longitudinally. In spite of the technical difficulties of testing such ideas, there has recently been one experimental demonstration of the circumferential recruitment of IDAs at the flagellar base [25]. The importance of circumferential recruitment at the base is that it would re-initiate itself and thus explain cyclical beating.

For simplicity, and to conform with the assumptions of the sliding-doublet theory, helical waveforms can be modelled without involving axonemal torsion. However, a truly comprehensive biochemical and physiological theory of the axoneme will need to encompass various three-dimensional phenomena. Two indications of the potential that axonemes have for three-dimensional motion may be found in recent descriptions from my laboratory of torque development by some sperm flagella. These are the alternating torsions of the axoneme associated normally with the propagation of a three-dimensional bend pair but observable also in the absence of perceptible bending [26]. This effect was interpreted as being due to simultaneous activation of all nine dynein arrays in a propagating zone giving local torsion and leading to local tension development followed by a chiral deformation. The second example of torque development is the mysteriously developed torque that can cause a virtually straight axoneme to spin about its longitudinal axis at frequencies exceeding 90 Hz [27]. These two examples recall the earlier evidence that the CP is induced to spin during axonemal beating [28], a finding that still awaits an explanation.

Summary

- *Axonemal dyneins occur in two rows (as inner and outer arms) on each of the nine doublets. Axonemal dynein binds reversibly to the B-microtubule and has an ATP-insensitive anchorage to the A-microtubule of the adjacent doublet.*
- *The heavy chains have the form of globular heads and are responsible for chemo-mechanical transduction. The B-tubule-binding site is on a tenuous extension of the head.*

- *There is only one type of ODA. A 12 nm shift in the globular heads is associated with the hydrolysis cycle.*
- *There are three types of IDA. No functional changes have been recognized in their complex conformation.*
- *There is plentiful evidence that the axonemal dyneins produce inter-doublet displacement. Doubt remains on how much sliding occurs per cycle of ATP hydrolysis. The mechanism for transforming sliding into bending is not yet explained.*

Work in my laboratory has been supported by the Wellcome Trust and the BBSRC (U.K.). I thank Dr. Geraint Vernon for suggesting improvements to an earlier draft of this chapter.

References

1. Afzelius, B.A., Dallai, R., Lanzavecchia, S. & Bellon, P.L. (1995) Flagellar structure in normal human spermatozoa and in spermatozoa that lack dynein arms. *Tissue Cell* **27**, 241–247
2. Gibbons, I.R. (1963) Studies on the protein components of cilia from *Tetrahymena pyriformis*. *Proc. Natl. Acad. Sci. U.S.A.* **50**, 1002–1010
3. Mitchell, D.R. (1994) Cell and molecular biology of flagellar dyneins. *Int. Rev. Cytol.* **155**, 141–180
4. Gibbons, I.R. (1998) Molecular biology of axonemal dynein and its role in ciliary motility. In *Cilia, Mucus, and Muco-ciliary Interactions* (Baum, G.L. et. al., eds.), pp. 1–11, Marcel Dekker, New York
5. Gee, M.A., Heuser, J.E. & Vallee, R.B. (1997) An extended microtubule-binding structure within the dynein motor domain. *Nature (London)* **390**, 636–639
6. King, S.M., Patel-King, R.S., Wilkerson, C.G. & Witman, G.B. (1995) The 78,000 M_r intermediate chain of *Chlamydomonas* outer arm dynein is a microtubule-binding protein. *J. Cell Biol.* **131**, 399–409
7. King, S.M. & Patel-King, R.S. (1995) The M_r=8000 and 11,000 outer arm dynein light chains from *Chlamydomonas* flagella have cytoplasmic homologues. *J. Biol. Chem.* **270**, 11445–11452
8. Burgess, S.A. (1995) Rigor and relaxed outer dynein arms in replicas of cryofixed motile flagella. *J. Mol. Biol.* **250**, 52–63
9. Sakakibara, H. & Nakayama, H. (1998) Translocation of microtubules caused by the αβ and γ outer arm dynein subparticles of *Chlamydomonas*. *J. Cell Sci.* **111**, 1155–1164
10. Vale, R.D. & Toyoshima, Y.Y. (1988) Rotation and translocation of microtubules *in vitro* induced by dyneins from *Tetrahymena* cilia. *Cell* **52**, 459–469
11. Goodenough, U.W. & Heuser, J.E. (1989) Structure of the soluble and *in situ* ciliary dyneins visualised by quick-frozen deep-etch microscopy. In *Cell Movement vol. 1, The Dynein ATPases* (Warner, F.D. et al., eds.), pp. 121–140, Alan R. Liss, New York
12. Kagami, O. & Kamiya, R. (1992) Translocation and rotation of microtubules caused by multiple species of *Chlamydomonas* inner-arm dynein. *J. Cell Sci.* **103**, 653–664
13. Woolley, D.M. (1998) Studies on the eel sperm flagellum. 1. The structure of the inner dynein arm complex. *J. Cell Sci.* **110**, 85–94
14. Kamiya, R. (1995) Exploring the function of inner and outer dynein arms with *Chlamydomonas* mutants. *Cell Motil. Cytoskeleton* **32**, 98–102
15. Bozkurt, H.H. & Woolley, D.M. (1993) Morphology of nexin links in relation to interdoublet sliding in the sperm flagellum. *Cell Motil. Cytoskeleton* **24**, 109–118
16. Woolley, D.M. (1998) Studies on the eel sperm flagellum. 2. The kinematics of normal motility. *Cell Motil. Cytoskeleton* **39**, 233–245

17. Brokaw, C.J. (1991) Microtubule sliding in swimming sperm flagella: direct and indirect measurements on sea urchin and tunicate spermatozoa. *J. Cell Biol.* **114**, 1201–1215

18. Satir, P. (1985) Switching mechanisms in the control of ciliary motility. *Modern Cell Biol.* **4**, 1–46

19. Gibbons, I.R. (1982) Sliding and bending in sea urchin sperm flagella. *Symp. Soc. Exp. Biol.* **35**, 227–287

20. Vernon, G.G. & Woolley, D.M. (1995) The propagation of a zone of activation along groups of flagellar doublet microtubules. *Exp. Cell Res.* **220**, 482–494

21. Johnson, K.A. (1985) Pathway of the microtubule-dynein ATPase and the structure of dyneins: a comparison with actomyosin. *Ann. Rev. Biophys. Biophys. Chem.* **14**, 161–188

22. Satir, P. (1989) Structural analysis of the dynein cross-bridge cycle. In *Cell Movement vol. 1. The Dynein ATPases* (Warner, F.D. et al., eds.), pp. 219–234, Alan R. Liss, New York

23. Shingyoji, C., Higuchi, H., Yoshimura, M., Katayama, E. & Yanagida, T. (1998) Dynein arms are oscillating force generators. *Nature (London)* **393**, 711–714

24. Ishijima, S., Kubo-Irie, M., Mohri, H. & Hamaguchi, Y. (1996) Calcium-dependent bidirectional power stroke of the dynein arms in sea urchin sperm axonemes. *J. Cell Sci.* **109**, 2833–2842

25. Woolley, D.M. (1998) Studies on the eel sperm flagellum. 3. Vibratile motility and rotatory bending. *Cell Motil. Cytoskeleton* **39**, 246–255

26. Woolley, D.M. & Vernon, G.G. (1999) Alternating torsions in a living 9+2 flagellum. *Proc. R. Soc. Lond. B* **266**, 1271–1275

27. Vernon, G.G. & Woolley, D.M. (1999) Three-dimensional motion of avian spermatozoa. *Cell Motil. Cytoskeleton* **42**, 149–161

28. Omoto, C.K. & Kung, C. (1980) Rotation and twist of the central-pair microtubules in the cilia of *Paramecium*. *J. Cell Biol.* **103**, 1895–1902

10

Translational elongation factor G: a GTP-driven motor of the ribosome

Wolfgang Wintermeyer[1] & Marina V. Rodnina

Institute of Molecular Biology, University of Witten/Herdecke, 58448 Witten, Germany

Introduction

Classical protein motors move along cytoskeletal filaments to produce movement within the cell or of the cell relative to its surroundings at the expense of ATP hydrolysis. Examples include myosin on actin or kinesin on microtubules. However, there is also energy-dependent, directional molecular movement in systems not involved in macroscopic cellular movement. Examples are enzymes that move in one direction along a nucleic acid, either changing its structure, such as DNA and RNA helicases driven by ATP hydrolysis, or using it as a template, for instance polymerases that synthesize DNA or RNA from nucleoside triphosphates on a DNA template. Among the latter, RNA polymerase is particularly interesting; the force developed by single molecules of *Escherichia coli* RNA polymerase has been measured, thereby directly demonstrating the mechanochemical function of the enzyme [1].

Another example, which is less well-established among biological motors, is the translating ribosome. During the elongation phase of protein synthesis, the ribosome polymerizes amino acids from amino acyl-tRNAs (aa-tRNAs) directed by an mRNA template that is moved codon by codon in successive

[1]*To whom correspondence should be addressed.*

rounds of elongation (Figure 1). In the first step, aa-tRNA is bound to the A site (the aa-tRNA-binding site). This step is catalysed by elongation factor Tu (EF-Tu), which forms a tight complex with aa-tRNA and GTP. Following codon recognition in the A site, EF-Tu hydrolyses GTP, undergoes a large conformational change, and releases aa-tRNA into the peptidyltransferase centre. Peptide-bond formation follows rapidly. Each time a new peptide bond has been formed, the mRNA together with peptidyl-tRNA and deacylated tRNA bound to it is moved along the ribosome by one codon (translocation). After this movement, peptidyl-tRNA resides in the P site (the peptidyl-tRNA-binding site) and deacylated tRNA is bound in a labile state to the exit site (E site) from where it dissociates spontaneously. In bacteria, translocation is promoted by elongation factor G (EF-G); in eukaryotes, the functional and structural homologue of EF-G is elongation factor 2 (EF-2).

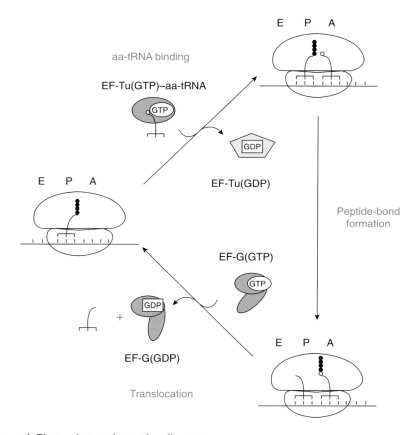

Figure 1. Elongation cycle on the ribosome

This simplified scheme depicts the major steps of elongation. Upon binding of the complex of aa-tRNA with elongation factor Tu (EF-Tu) and GTP to the ribosomal A site and codon recognition, GTP is hydrolysed, EF-Tu(GDP) is released, and aa-tRNA enters the A site to take part in peptide-bond formation. Subsequent translocation entails binding of EF-G(GTP), GTP hydrolysis, tRNA–mRNA movement and release of EF-G(GDP) and deacylated tRNA.

EF-G, EF-Tu and several other translation factors form a subgroup of the GTPase superfamily. The members of this subgroup are large GTPases that contain, in addition to the GTP-binding domain, other functional domains. Canonical GTPases, such as p21 Ras or the G_α subunit of heterotrimeric G-proteins, act as molecular switches that are 'turned on' by replacing GDP with GTP to assume the GTP-bound, active conformation, and are 'switched off' by GTP hydrolysis. In the active conformation they can bind other proteins (effectors) with high affinity. It was thought that EF-G functions in a similar fashion, but recently it was found that EF-G is active in its GDP-bound state. Thus EF-G appears to differ from canonical GTPases; rather, it resembles motor proteins, such as the myosin motor, which performs the force-generating conformational change (power stroke) after ATP hydrolysis and actin binding.

This chapter discusses the role of EF-G and GTP hydrolysis in translocation on the ribosome. EF-G from *E. coli* will be used to discuss function; for structural information, we refer to the highly homologous factor from *Thermus thermophilus*. The main emphasis will be on how EF-G may translate the free energy of GTP hydrolysis into molecular movement on the ribosome.

Structure of EF-G: a five-domain GTPase with unusual properties

EF-G from *E. coli* is a 77 kDa protein with five domains. Whereas the structure of the GTP-bound form is not known, the crystal structures of nucleotide-free and GDP-bound EF-G forms from *T. thermophilus* have been determined and are quite similar [2–4]. Figure 2 shows the structure of EF-G(GDP) at 2.6 Å resolution. The molecule is elongated (about 110 Å long) and consists of a body (domains 1–3) to which an extended arm (domains 4 and 5) is connected through domains 3 and 5. The G domain (domain 1) of EF-G is larger (293 amino acids) than in other GTPases, such as p21 Ras or EF-Tu (about 200 amino acids), due to the insertion of a subdomain, G' (90 amino acids), into the structurally conserved G-domain framework.

This structural difference in the G domain is reflected in the unusual nucleotide-binding properties of EF-G; that is, its low affinities for GTP (1 μM) and GDP (10 μM), and the ease of GDP dissociation which obviates the necessity for a nucleotide-exchange factor. On the basis of these properties one may expect that the structures of the GDP-bound form and the as-yet-unknown GTP-bound form of EF-G would be similar, in contrast to EF-Tu where the structural change due to GTP hydrolysis is very large [5,6] and an exchange factor (elongation factor Ts, EF-Ts) is required to catalyse the dissociation of GDP. Results from small-angle X-ray scattering experiments indeed indicate that the difference in the overall structures of GTP and GDP forms of EF-G in solution is small [7].

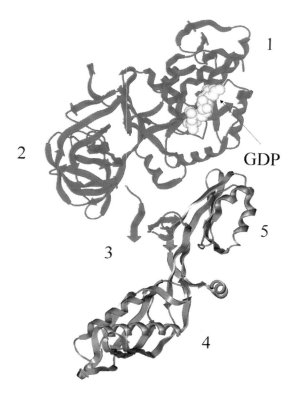

Figure 2. Structure of EF-G(GDP)
Crystal structure of EF-G(GDP) from *T. thermophilus* at 2.6 Å resolution [4]. Domains are indicated by numbers; GDP is shown in domain 1 (G domain). Domain 3 is not well-defined in the crystal, indicating flexibility in this domain.

This may not be true for the ribosome-bound factor, where the nucleotide-binding properties of EF-G may be different and the factor may experience more extensive conformational changes depending on the ligands bound to the G domain, i.e. GTP, GDP and P_i, or GDP. Binding experiments with non-hydrolysable analogues of GTP or GDP indeed show that nucleotides are bound more tightly by EF-G on the ribosome, thus meeting one prerequisite for GTP hydrolysis to generate force on the ribosome. There is, however, no direct information available to characterize those states.

By analogy with the well-characterized structural change in the G domain of EF-Tu brought about by GTP hydrolysis [5,6], it appears likely that GTP hydrolysis and the loss of the γ-phosphate induces a rearrangement in the G domain of EF-G. This may change the overall structure of the molecule by affecting the interaction of the G domain with neighbouring domains. It is appealing to assume that this change results in a movement of domain 4 in a lever-like fashion. It may be induced by a change in the interaction of domain 1 with domain 5, which forms a structural unit with domain 4, with domain 3 serving as a hinge. The interactions of domain 1 with domain 5 (and 2) do seem to be functionally important, since numerous fusidic acid-resistance mutations and revertants have been found at the domain interfaces [8], suggesting that fusidic acid, by binding to the interfaces, inhibits intramolecular rearrangements. Further support comes from the finding that deletion of domain 1 abolishes the ability of EF-G to promote translocation on the small ribosomal subunit [9].

Mechanism of EF-G-dependent translocation

Until recently, the function of EF-G was explained in terms of the classical GTPase paradigm in which EF-G(GTP) binds to the pre-translocation ribosome and induces a conformational change that allows translocation and, subsequently, EF-G hydrolyses GTP, switches to a low-affinity GDP-bound conformation, and dissociates from the ribosome. The model was based on the observation that EF-G with non-hydrolysable GTP analogues still enhances translocation, but is restricted to a single round, unless the factor is actively removed from the ribosome after translocation [10]. However, on the basis of the kinetic data discussed below, we have recently proposed the revised mechanism of EF-G function in translocation, depicted in Figure 3.

In the pre-translocation complex, deacylated tRNA resides in the P site (or, more precisely, in the P/E hybrid state [11]) and peptidyl-tRNA in the A site (A/P* state; P* indicating a partially translocated state). Immediately following the binding of EF-G(GTP) (step 1), GTP is hydrolysed (step 2). Most likely, GTP hydrolysis and/or subsequent release of P_i causes a conformational change of EF-G which, in turn, induces the formation of the transition state

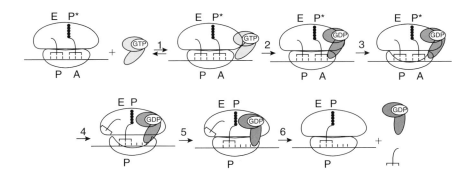

Figure 3. Reaction scheme of EF-G-dependent translocation
EF-G is depicted in three different conformations: the GTP-bound form, an intermediate GDP-bound form in the pre-translocation state, and the GDP-bound form in the post-translocation state that dissociates from the ribosome (compare with Figure 2). P_i is omitted as it is not known when it is released. The transition-state structure of the ribosome, formed in step 3, is symbolized by an altered conformation of the small subunit. A, P (P* indicating a partially translocated state) and E denote the tRNA-binding sites on the two ribosomal subunits; only occupied sites are indicated. Reprinted from [22], with permission. ©1998, National Academy of Sciences, U.S.A.

of the ribosome (step 3). In the transition state, the movement of the tRNA–mRNA complex takes place (step 4). In step 5, the ribosome returns to the ground state and EF-G assumes the GDP-bound conformation (this is probably the step inhibited by fusidic acid binding to EF-G on the ribosome). Step 6 then comprises the dissociation of EF-G(GDP) and deacylated tRNA (order unknown) to reach the final post-translation state of the ribosome with peptidyl-tRNA in the P site (P/P state) and a free A site.

That GTP hydrolysis is the first event after binding of EF-G(GTP) to the ribosome follows from pre-steady-state kinetic experiments [12] that clearly showed that EF-G hydrolyses GTP much faster than the translocation of peptidyl-tRNA takes place (Figure 4). The rate constant of single-round GTP hydrolysis is about 170 s^{-1} at saturating concentrations of EF-G, whereas the rate of peptidyl-tRNA movement is 25 s^{-1}. There is one molecule of GTP hydrolysed during each translocation cycle. The rate of GTP hydrolysis is

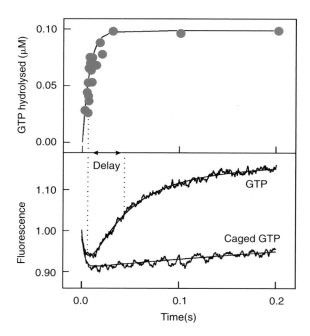

Figure 4. Kinetics of EF-G-dependent GTP hydrolysis by EF-G and of translocation
The upper graph shows the time course of [γ-^{32}P]GTP hydrolysis measured at 37°C by quench-flow at 0.1 μM pre-translocation ribosomes and 2 μM EF-G. Multiple turnover hydrolysis of [γ-^{32}P]GTP (5 μM, added with EF-G) following translocation (uncoupled GTP hydrolysis) was suppressed by adding unlabelled GTP (1 mM) with the ribosomes. The lower graph shows translocation measured by fluorescence stopped-flow. The fluorescence signal of proflavin-labelled peptidyl-tRNA was monitored in the presence of either GTP or caged GTP (both 1 mM). The initial fluorescence decrease is due to EF-G binding, the slower increase to translocation. Single- or two-exponential fits are depicted by smooth lines. The time delay between GTP hydrolysis and translocation is indicated. Reprinted with permission from *Nature* [12] ©1997 Macmillan Magazines Limited.

independent of the status of the ribosome with respect to the occupancy with tRNA and is not affected when translocation is blocked by the antibiotic viomycin [12]. Thus single-round GTP hydrolysis by EF-G on the ribosome does not depend upon translocation.

GTP hydrolysis is required for rapid translocation. This is shown directly by the observation that GTP hydrolysis accelerates translocation more than 50 times relative to the reaction with the non-hydrolysable GTP analogue, caged GTP (Figure 4, lower panel). Accordingly, there seem to be two ways by which EF-G catalyses translocation. In one, operating with non-hydrolysable GTP analogues or GDP [12], the activation energy of translocation is reduced by the binding of EF-G to the pre-translocation complex, which is not very much affected by the nucleotide. Subsequent translocation is relatively slow and the turnover of EF-G even slower, because the transition state is reached by thermal motion only. By contrast, in the mechanism operating in the presence of GTP, the hydrolysis of GTP results in a substantial acceleration. From the rates of GTP hydrolysis and translocation, a time gap between the two reactions of about 35 ms is calculated (Figure 4). The lag suggests that conformational strain introduced by GTP hydrolysis is stored in the complex and released by proceeding to the transition state and translocation.

The steps from GTP hydrolysis on formation of the transition state, translocation and release of EF-G depend on the presence of domain 4 of EF-G [12]. The deletion of domain 4 reduced the activity in translocation about 1000-fold, while it had no effect on the GTPase activity of EF-G on the ribosome. Most interestingly, both activities were restricted to a single round, indicating tight binding of the truncated factor to the ribosome (Figure 5). Mutant EF-G lacking both domains 4 and 5 behaved the same. The data suggest that domain 4 is essential for (i) coupling the conformational change of EF-G induced by GTP hydrolysis to structural rearrangements of the ribosome leading to tRNA translocation, as well as for (ii) subsequent release of the factor.

The striking observation that truncated EF-G lacking domain 4, or domains 4 and 5, does not dissociate from the ribosome following GTP hydrolysis suggests that, early in the sequence of conformational changes induced by GTP hydrolysis, additional interactions of EF-G with the ribosome are established to form a kinetically stable complex, and that these interactions involve the body of EF-G, that is, domains 1–3. As a consequence, there is conformational coupling between the factor and the ribosome, which enables the factor to force the ribosome into the transition state of translocation. The structure of the transition-state complex revealed by electron cryomicroscopy, discussed below, suggests that domain 4 functions like a lever arm that acts on the 30 S ribosomal subunit, thereby inducing a structural transition. Subsequent tRNA–mRNA movement is likely to be spontaneous, following the thermodynamic gradient towards the immediate post-translocation P site–E site arrangement of the tRNAs, although an active contribution of EF-G is not excluded [13]. A movement of domain 4 towards or into the vac-

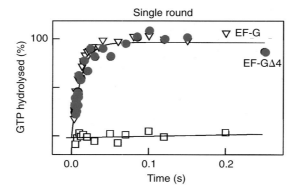

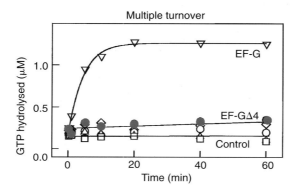

Figure 5. Effect of domain-4 deletion on GTP hydrolysis by EF-G on the ribosome
The upper graph shows a single round of GTP hydrolysis with 0.8 μM EF-G (▽) or EF-GΔ4 (EF-G domain 4, ●), in each case plus 0.2 μM pre-translocation complex and 5 μM [γ-^{32}P]GTP at 37°C. Control (□) is minus EF-G. The lower graph shows multiple rounds with 0.3 μM EF-G (▽) or EF-GΔ4 (●), and other reagents as in single-round hydrolysis. Controls were run with ribosomes alone (◇), EF-G alone (△) and EF-GΔ4 alone (○). Reprinted with permission from *Nature* [12] ©1997 Macmillan Magazines Limited.

ated 30 S A site (see below) seems to be required to resolve the tight interaction. The mutant lacking domain 4 cannot perform this movement and, therefore, retains the high affinity gained by GTP hydrolysis, thereby preventing turnover.

Ribosomal movements in translocation

On the 30 S subunit of the ribosome, the anti-codon regions of the two tRNAs are bound to two contiguous codons of the mRNA in the decoding centre, which is formed by residues 1400–1410 and 1490–1501 of 16 S rRNA. Binding interactions between some of these residues and the anti-codon arms of both tRNAs contribute to the stabilization of the complex. For tRNA–mRNA movement during translocation, one may envisage two principal possibilities. One is that there is a large-scale movement of the part of the ribosome which binds the tRNAs, thus carrying the tRNA–mRNA complex along. This type of movement is probably excluded because the tRNA–16 S rRNA contacts during the movement change from the typical pre-translocation pattern to the post-translocation pattern [11], indicating that the tRNAs move relative to the ribosome. The alternative, more probable, mechanism is that the binding interactions between 16 S rRNA and the tRNA–mRNA complex are released prior to movement and are reformed afterwards with a new tRNA partner (peptidyl-tRNA) in the P site and no tRNA in the A site (the deacylated tRNA, now bound in the E site, does not make strong contacts with the 30 S subunit).

Both models require that the conformation of the ribosome, in particular of the 30 S subunit, changes to allow translocation. Indeed, in the pre-translocation state with EF-G bound to it, the ribosome does have a structure, revealed by electron cryomicroscopy, that is substantially different from the structures of both the initial EF-G-free state and the final post-translocation state, which are very similar (Stark, H., Rodnina, M.V., van Heel, M. & Wintermeyer, W., unpublished work). The most extensive structural differences are found on the 30 S subunit. These results suggest that translocation is indeed associated with a major conformational change of the ribosome, and we assume that the state visualized by electron cryomicroscopy represents, or is closely related to, the transition state of translocation.

In the transition state, EF-G is arranged across the intersubunit cleft such that the body of EF-G is oriented towards the 50 S subunit, making a strong contact with ribosomal proteins L7 and L12, while the tip of domain 4 contacts the 30 S subunit in the region where protein S4 is located. This arrangement of EF-G suggests how EF-G may induce the structural change of the ribosome leading to translocation. We postulate that EF-G, with its body firmly bound elsewhere on the ribosome, imposes a conformational strain on the 30 S subunit by an active movement of domain 4 bound to the 30 S subunit, thereby initiating a structural change of 16 S rRNA that is transmitted to the decoding centre. In turn, the conformational change in the decoding centre may result in a destabilization of 16 S rRNA–tRNA interactions, which allows the tRNA–mRNA movement.

After translocation, EF-G is found in a different position on the ribosome, domain 4 now reaching into the decoding centre. This arrangement was suggested by chemical data [15] and demonstrated directly by electron cryomicroscopy (Stark, H., Rodnina, M.V., van Heel, M. & Wintermeyer, W., unpublished work; [16]). While the reorientation of EF-G may be related to the movement of the tRNA–mRNA complex, it seems to be essential for the decay of the tight complex after translocation, as discussed above. It is noteworthy that the arrangement of EF-G in the post-translocation complex is quite similar to the arrangement of EF-Tu–aa-tRNA in the codon-recognition complex, where the anti-codon arm reaches into the decoding centre [17]. This is in keeping with the strikingly similar tertiary structures of EF-G and the EF-Tu–aa-tRNA complex in which domain 4 of EF-G and the anti-codon arm of the tRNA match each other in shape and position [18].

EF-G: a GTPase motor

Several features of the functional cycle of EF-G are inconsistent with a GTPase switch model: (i) GTP is hydrolysed early in the cycle, immediately following the binding of EF-G(GTP) to the ribosome; (ii) after GTP hydrolysis, the ribosome–EF-G(GDP) complex rearranges into a kinetically stable state; (iii) there is a time delay, about 35 ms, between GTP hydrolysis and the actual tRNA movement, indicating storage of conformational strain in the ribosome-factor complex; and (iv) domain 4 of EF-G appears to function as a lever arm exerting conformational strain on the ribosome.

These characteristics of EF-G function have interesting parallels in myosin function. The basic cycle of myosin action on actin starts with the ATP-bound form of the myosin head detached from actin. Subsequent ATP hydrolysis is spontaneous, and the myosin head with ADP and P_i bound to it attaches to actin to form a reversible complex. On release of P_i, the complex rearranges into a tight complex that then performs the power stroke, thus releasing the conformational strain stored in the tight complex. Finally, a structural transition in the myosin head, induced by dissociation of ADP and binding of ATP, results in the detachment of the myosin head from actin.

Common structural themes of G-proteins and molecular motors have been discussed recently [19]. Whereas there are striking parallels, both structural and functional, in the nucleotide-binding domains and in the structural changes induced by nucleoside triphosphate hydrolysis, there are important differences in the consequences of nucleoside triphosphate hydrolysis. GTPases are active generally in the GTP-bound form, and the structural change in the G domain induced by GTP hydrolysis destabilizes interactions with effectors. In contrast, ATP hydrolysis in molecular motors promotes the binding to partner molecules, for instance of myosin to actin and, additionally, changes the overall architecture of the molecule, thereby introducing conformational strain. Given the appropriate molecular environment, the conformational strain can exert force on the interaction partner. The characteristics of EF-G function, in particular the intermediate formation of a tight ribosome–elongation-factor complex in which conformational strain induces a structural transition in the ribosome, are consistent with the latter scenario. Although molecular details of myosin function are being successfully revealed by structural analysis [20], not much is yet known for EF-G, as the structures of only the GDP-bound and nucleotide-free forms (but not the GTP-bound form) have been determined.

Other force-generating GTPases related to EF-G

In eukaryotes, translocation on the ribosome is promoted by EF-2, which shares extensive homologies with EF-G. Thus while less mechanistic information is available for EF-2, it is very likely that it functions in the same way as EF-G. Interestingly, an evolutionarily conserved GTPase homologous to EF-2 has been found as a constituent of human and yeast spliceosomes, i.e. U5 snRNP (small nuclear ribonuclear protein), and shown to be essential for yeast cell viability [21]. Thus the possibility arises that in addition to structural rearrangements driven by ATP hydrolysis, the RNA splicing machine may undergo a transition promoted by GTP hydrolysis that may be analogous to the structural transition of the ribosome that leads to translocation.

Perspectives

While the model of EF-G function described here provides a consistent explanation of the available data, it is partly hypothetical, and many questions are still open. Most importantly, the structural changes implied for EF-G are speculative. Thus it will be very important to obtain the crystal structure of EF-G(GTP) but it should be remembered that comparing the crystal structures of free EF-G(GTP) and EF-G(GDP) may not necessarily tell us about the structure in the most important functional state, bound to the ribosome in the transition state of translocation. However, we may get an idea of this from high-resolution electron cryomicroscopy of the ribosome–elongation-factor complex at defined stages of translocation.

Another unknown is how the binding of nucleotide, and possibly of P_i, to EF-G is stabilized on the ribosome, since GDP–GTP exchange before completion of translocation would probably interfere with maintaining the conformational strain in EF-G generated by GTP hydrolysis. It is conceivable that a ribosomal component which contacts the G domain of EF-G and triggers GTP hydrolysis (a likely candidate is protein L7/L12) closes the nucleotide-binding pocket. Thereby, the release of GDP and/or P_i could be inhibited until further interactions with the ribosome are established which lead to reopening of the pocket. To resolve these questions, the interactions of EF-G with the ribosome have to be characterized in molecular detail, and the timing of P_i release has to be measured.

Summary

- *EF-G is a large, five-domain GTPase that promotes the directional movement of mRNA and tRNAs on the ribosome in a GTP-dependent manner.*
- *Unlike other GTPases, but by analogy to the myosin motor, EF-G performs its function of powering translocation in the GDP-bound form; that is, in a kinetically stable ribosome–EF-G(GDP) complex formed*

by GTP hydrolysis on the ribosome. The complex undergoes an extensive structural rearrangement, in particular affecting the small ribosomal subunit, which leads to mRNA–tRNA movement. Domain 4, which extends from the 'body' of the EF-G molecule much like a lever arm, appears to be essential for the structural transition to take place.

- *In a hypothetical model, GTP hydrolysis induces a conformational change in the G domain of EF-G which affects the interactions with neighbouring domains within EF-G. The resulting rearrangement of the domains relative to each other generates conformational strain in the ribosome to which EF-G is fixed. Because of structural features of the tRNA–ribosome complex, this conformational strain results in directional tRNA–mRNA movement.*

- *The functional parallels between EF-G and motor proteins suggest that EF-G differs from classical G-proteins in that it functions as a force-generating mechanochemical device rather than a conformational switch. There are other multi-domain GTPases that may function in a similar way.*

The work was supported by the Deutsche Forschungsgemeinschaft, the Alfried Krupp von Bohlen und Halbach-Stiftung and the Fonds der Chemischen Industrie.

References

1. Yin, H., Wang, M.D., Svoboda, K., Landick, R., Block, S.M. & Gelles, J. (1995) Transcription against an applied force. *Science* **270**, 1653–1657

2. Czworkowski, J., Wang, J., Steitz, T.A. & Moore, P.B. (1994) The crystal structure of elongation factor G complexed with GDP, at 2.7 Å resolution. *EMBO J.* **13**, 3661–3668

3. Ævarsson, A., Brazhnikov, E., Garber, M., Zheltonosova, J., Chirgadze, Y., al-Karadaghi, S., Svensson, L.A. & Liljas, A. (1994) Three-dimensional structure of the ribosomal translocase: elongation factor G from *Thermus thermophilus. EMBO J.* **13**, 3669–3677

4. al-Karadaghi, S., Ævarsson, A., Garber, M., Zheltonosova, J. & Liljas, A. (1996) The structure of elongation factor G in complex with GDP: conformational flexibility and nucleotide exchange. *Structure* **4**, 555–565

5. Polekhina, G., Thirup, S., Kjeldgaard, M., Nissen, P., Lippmann, C. & Nyborg, J. (1996) Helix unwinding in the effector region of elongation factor EF-Tu-GDP. *Structure* **4**, 1141–1151

6. Abel, K., Yoder, M.D., Hilgenfeld, R. & Jurnak, F. (1996) An α to β conformational switch in EF-Tu. *Structure* **4**, 1153–1159

7. Czworkowski, J. & Moore, P.B. (1997) The conformational properties of elongation factor G and the mechanism of translocation. *Biochemistry* **36**, 10327–10334

8. Johanson, U., Ævarsson, A., Liljas, A. & Hughes, D. (1996) The dynamic structure of EF-G studied by fusidic acid resistance and internal revertants. *J. Mol. Biol.* **258**, 420–432

9. Borowski, C., Rodnina, M.V. & Wintermeyer, W. (1996) Truncated elongation factor G lacking the G domain promotes translocation of the 3' end but not of the anticodon domain of peptidyl-tRNA. *Proc. Natl. Acad. Sci. U.S.A.* **93**, 4202–4206

10. Spirin, A.S. (1985) Ribosomal translocation: facts and models. *Progr. Nucleic Acid. Res. Mol. Biol.* **32**, 75–114

11. Moazed, D. & Noller, H.F. (1989) Intermediate states in the movement of transfer RNA in the ribosome. *Nature (London)* **342**, 142–148

12. Rodnina, M.V., Savelsbergh, A., Katunin, V.I. & Wintermeyer, W. (1997) Hydrolysis of GTP by elongation factor G drives tRNA movement on the ribosome. *Nature (London)* **385**, 37–41

13. Abel, K. & Jurnak, F. (1996) A complex profile of protein elongation: translating chemical energy into molecular movement. *Structure* **4**, 229–238

14. Reference deleted.

15. Wilson, K.S. & Noller, H.F. (1998) Mapping the position of translational elongation factor EF-G in the ribosome by directed hydroxyl radical probing. *Cell* **92**, 131–139

16. Agrawal, R.K., Penczek, P., Grassucci, R.A. & Frank, J. (1998) Visualization of elongation factor G on the *Escherichia coli* 70 S ribosome: the mechanism of translocation. *Proc. Natl. Acad. Sci. U.S.A.* **95**, 6134–6138

17. Stark, H., Rodnina, M.V., Rinke-Appel, J., Brimacombe, R., Wintermeyer, W. & van Heel, M. (1997) Visualization of elongation factor Tu on the *Escherichia coli* ribosomes. *Nature (London)* **389**, 403–406

18. Nissen, P., Kjeldgaard, M., Thirup, S., Polekhina, G., Reshetnikova, L., Clark, B.F. & Nyborg, J. (1995) Crystal structure of the ternary complex of Phe-tRNA^{Phe}, EF-Tu, and a GTP analog. *Science* **270**, 1464–1472

19. Vale, R.D. (1996) Switches, latches, and amplifiers: Common themes of G proteins and molecular motors. *J. Cell Biol.* **135**, 291–302

20. Dominguez, R., Freyzon, Y., Trybus, K.M. & Cohen, C. (1998) Crystal structure of a vertebrate smooth muscle myosin motor domain and its complex with the essential light chain: visualization of the pre-power stroke state. *Cell* **94**, 559–571

21. Fabrizio, P., Laggerbauer, B., Lauber, J., Lane, W.S. & Lührmann, R. (1997) An evolutionary conserved U5 snRNP-specific protein is a GTP-binding factor closely related to the ribosomal translocase EF-2. *EMBO J.* **16**, 4092–4106

22. Rodnina, M.V. & Wintermeyer, W. (1998) Form follows function: structure of an elongation factor G-ribosome complex. *Proc. Natl. Acad. Sci. U.S.A.* **95**, 7237–7239

11

How do proteins move along DNA? Lessons from type-I and type-III restriction endonucleases

Mark D. Szczelkun

Department of Biochemistry, University of Bristol, Bristol BS8 1TD, U.K.

Introduction

A fundamental requirement of nearly every genetic event, including DNA replication, transcription, cleavage and repair, is the protein-mediated communication between distant DNA sites. Four general mechanisms for this 'action at a distance' were described by Adzuma and Mizuuchi [1] and are illustrated in Figure 1. The first three models, DNA sliding, DNA hopping and DNA looping, are driven passively by random thermal fluctuations of the protein or DNA, with significant consequences for the efficiency of the reactions. In DNA sliding (Figure 1a), a protein bound to non-specific DNA diffuses either leftwards or rightwards without dissociating, altering the DNA-binding register by 1 bp. By confining diffusion to one dimension, the probability of finding a second site or protein is increased. However, the number of physical steps needed to move between two DNA loci by linear diffusion has a square power dependence on the distance between those sites; for instance, travelling 1000 bp takes 1×10^6 steps, which in turn requires a linear diffusion rate at least 1×10^6 times faster than the DNA dissociation rate. Generally, DNA-binding proteins associate transiently with non-specific

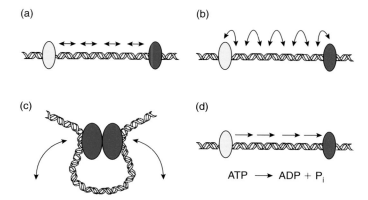

Figure 1. Protein-mediated communications between distant sites on DNA
DNA is shown as a helical ribbon. Proteins are represented as ovals; light blue corresponds to the starting position, dark blue to the finishing position. (a) DNA sliding. A protein bound to non-specific DNA randomly diffuses in one dimension without releasing the DNA. (b) DNA hopping (DNA dissociation-reassociation). A protein dissociates from the DNA, diffuses in three dimensions, then re-binds randomly at another non-specific DNA site. (c) DNA looping. A protein bound to a specific DNA site comes into close contact with a second DNA site or DNA-bound protein through the three-dimensional segmental diffusion of the DNA backbone. The binding of both sites sequesters the intervening DNA into a loop. (d) DNA translocation. A protein follows the one-dimensional DNA contour with a fixed polarity. Motion is driven by an external source of free energy, e.g. the hydrolysis of ATP.

sequences and DNA sliding would only be feasible over quite short distances, less than 100 bp. In DNA hopping (Figure 1b), a protein physically releases DNA, diffuses in three dimensions and randomly re-binds the same DNA molecule at an alternative site. The DNA acts as a binding 'reservoir' — given enough time, one protein could sample every available binding locus. Three-dimensional diffusion on a random coil has a square-root power dependence on the distance between target sites, but because sampling is non-linear, sites may be overlooked and the protein could even diffuse away from the DNA altogether. In DNA looping (Figure 1c), a protein remains bound to a specific DNA sequence whilst segmental diffusion of the polynucleotide brings distant DNA sites (and any bound proteins) into close proximity. In general, the tethering of two proteins on to the same DNA molecule increases the probability of protein–protein interactions. However, the probability of juxtaposition decreases as the distance between the sites increases, an effect that is particularly marked on linear DNA. DNA-looping events are less likely over distances greater than ≈1000 bp without the co-operation of accessory factors, such as DNA-bending proteins.

Many enzymes acting on nucleic acids must communicate efficiently over thousands of base pairs. For instance, single molecules of RNA polymerase faithfully transcribe entire genes. To achieve such a high degree of processivity requires an alternative strategy; DNA translocation (Figure 1d). Translocation resembles sliding in that the protein follows the one-dimensional DNA contour. But compared with linear diffusion, translocation occurs in one direction at a time, reducing significantly the number of steps required to move a given distance. To maintain polarity and processivity, movement must be coupled to an input of free energy. For instance, RNA polymerase uses free energy from the condensation of nucleoside triphosphates into the nascent RNA chain and the concomitant folding of the extruded transcript. Other enzymes, such as the DNA helicases, catalyse the hydrolysis of ATP or GTP. As such, the processive nucleic acid enzymes can be regarded as motor proteins — the conversion of chemical energy to mechanical motion is equivalent to that of the classical motor proteins reviewed elsewhere in this volume. In this chapter, the motion on DNA catalysed by the type-I and type-III restriction endonucleases will be reviewed, and related to other DNA-based motor proteins such as the DNA helicases.

Restriction/modification systems

The restriction/modification enzymes are distributed widely in bacteria, where they constitute the prokaryotic equivalent of an immune system, capable of discriminating between foreign and host DNA. This is achieved by two enzyme activities; an endonuclease and a methyltransferase. The endonuclease cleaves DNA phosphodiester bonds upon binding a specific sequence of nucleotides. The host DNA is protected from self-digestion by the methyltransferase, which catalyses transfer of methyl groups from the donor *S*-adenosyl methionine to particular bases within the recognition site. Conversely, invasive DNA does not carry the correct pattern of methylated bases and will be cleaved by the endonuclease.

Type-I and type-III restriction endonucleases

Restriction/modification enzymes are classified into three types according to their genetics, subunit composition and biochemical activity. The type-I and type-III enzymes are large oligomeric proteins (300–400 kDa), which carry out both methylation and cleavage reactions within the same complex [2]. Type-I enzymes are encoded by three genes, *hsdS* (which is responsible for DNA recognition), *hsdM* (DNA methylation) and *hsdR* (DNA and ATP hydrolysis). The typical subunit composition is $HsdR_2HsdM_2HsdS_1$ [3]. The type-III enzymes are encoded by two genes, *mod* (responsible for DNA recognition and methylation) and *res* (DNA and ATP hydrolysis). The subunits are arranged Mod_2Res_2 [2]. Originally, the distribution of both types of enzyme was thought to be limited to strains of enterobacteria. However, due to the recent increase in available genome sequences, it has become apparent that these enzymes are in fact ubiquitous amongst bacterial and archaebacterial species.

Tracking, stalling and cleavage by type-I restriction endonucleases

The type-I restriction endonucleases recognize specific asymmetric, bipartite sequences (e.g. GAANNNNNNRTCG for *Eco*R124I, where N is any base and R is a purine). An unexpected observation is that ensuing DNA hydrolysis does not occur within the recognition site, as would be expected for an archetypal type-II restriction endonuclease such as *Eco*RI, but takes place at random loci that can be anywhere between 50 and 11000 bp away from the site [2]. The reaction requires both Mg^{2+} ions and ATP cofactors, and in some cases *S*-adenosyl methionine. The interaction between site-specific recognition and non-specific cleavage is provided by DNA tracking [4,5], a one-dimensional process driven by ATP hydrolysis (Figure 2a). Initially, an enzyme associates with both its DNA-recognition site (through the HsdS subunit) and the adjacent non-specific DNA (probably through the HsdR subunits; see below). Despite asymmetry in the binding of HsdS to DNA, the domain organization provides a pseudo-dyad symmetry for the assembly of HsdM and HsdR [6,7]. For clarity, the model in Figure 2(a) is only illustrated with a single HsdR. As ATP is hydrolysed, non-specific DNA is pulled past the enzyme. Since the recognition site is not released [2], an expanding loop of double-stranded DNA is extruded. The model can be extended to accommodate multiply associated HsdR subunits [3] or bi-directional translocation via a twin-loop model (see below).

The first indication that cleavage was linked to motion was that ATP was hydrolysed in large amounts both during and after DNA cleavage [8]. Further evidence came from electron microscopy. Type-I enzymes bound to circular DNA in the presence of ATP produced figure-of-eight structures indicative of the tracking model [5,8]. On linear DNA, the expected α-DNA structures were also observed [4,8]. An unequivocal demonstration of linear motion resulted from measuring cleavage of interlinked rings of DNA called catenanes [9]. Proteins that passively loop DNA (Figure 1c) simply require two DNA sites tethered closely in space. Since the local concentration of two sites partitioned on to separate rings of a catenane is equivalent to two sites on one DNA molecule, the probability of sites interacting by DNA looping is the same on either substrate [1,9]. Conversely, any process that follows the DNA contour between sites is doomed to failure if the loci are on separate DNA rings, even when linked as a catenane. Cleavage of plasmid DNA carrying one *Eco*R124I site occurred throughout the DNA [9–11]. However, when the DNA was partitioned into a catenane, cleavage occurred exclusively in the ring carrying the recognition site; the second ring could not be cleaved despite being held in proximity to the first [9]. This result cannot be accommodated by any mechanism that requires DNA looping, even as a pre-requisite to tracking [5], and is only satisfied by the tracking model described in Figure 2.

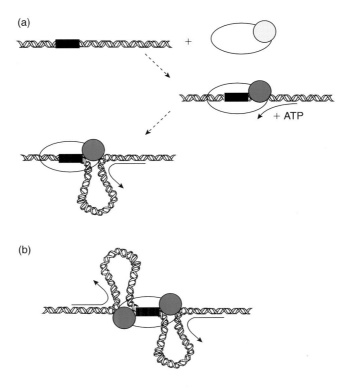

Figure 2. DNA tracking
DNA is shown as a helical ribbon with a specific recognition site shown as a black rectangle. For type-I enzymes, the large white oval represents one HsdS and two HsdM subunits, the blue circle represents either one or two HsdR subunits. For type-III enzymes, the white oval represents two Mod subunits and the blue circle represents two Res subunits. (a) Unidirectional model. The broken arrow represents multiple steps. The protein binds both to the specific recognition site and an adjacent non-specific segment. The protein remains bound at recognition site while the non-specific DNA is pulled past, dependent on the hydrolysis of ATP. The direction of tracking is only illustrative. (b) Bidirectional model. Tracking occurs from both sides of the enzyme simultaneously, extruding two DNA loops. See text for a full description of the models.

What causes DNA cleavage at distant sites? The simple answer is that DNA is cut wherever a pause in tracking occurs [9]. One way to stall motion is described by the collision model, first proposed by Studier and Bandyopadhyay to account for the distribution of cleavage sites produced on T7 phage DNA by *Eco*KI [12]. A series of DNA fragments was observed that corresponded to cleavage midway between each pair of neighbouring *Eco*KI sites. This was explained by assuming that when two enzymes track towards each other they will meet, on average, halfway between the two sites. Accordingly, T7 DNA was cleaved wherever two *Eco*KI enzymes collided (Figure 3a). The model also assumes that the direction of tracking is not predetermined by the relative orientations of the asymmetric recognition sites; this has also been confirmed for other type-I enzymes. Studier and Bandyo-

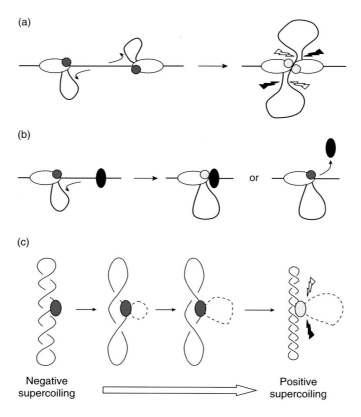

Figure 3. Mechanisms for stalling DNA tracking
DNA is represented as a black line, and proteins as ovals and circles. (a) DNA collision model. Two proteins translocate towards each other. When they collide (on average, half way between the sites), translocation stops. For the type-I enzymes, cleavage can occur either at distant loci (black flashes) or loci close to the recognition sites (white flashes). For the type-III enzymes, cleavage only occurs close to the recognition sites (white flashes). (b) Collision with a stationary DNA-binding protein (black oval) can have two outcomes; either the static protein is driven off of the DNA (type-I enzymes) or it causes the translocating enzyme to pause (type-III enzymes). (c) Tracking on circular DNA. Circular DNA is represented as a coiled line with either right-handed (negative) or left-handed (positive) supercoils. Unidirectional DNA tracking generates a figure-of-eight DNA structure. Increasing twist in the contracting loop reduces the negative writhe, until eventually positive writhe is introduced. When no more twist can be accommodated, the enzyme stalls and the DNA is cleaved (white or black flashes). The expanding loop is shown as a dotted line without any writhe; the decrease in twist cannot be accommodated and must be removed (see text for a full explanation).

padhyay therefore proposed that tracking must be bidirectional [12]; DNA is tracked from both sides of the recognition site, so extruding two DNA loops (Figure 2b). However, there is evidence that some type-I enzymes, such as *Eco*BI, can only cleave DNA on one side of their asymmetric recognition site [4], indicating that translocation has a fixed polarity (Figure 2a).

None of the type-I enzymes can cleave linear DNA carrying a single recognition site at stoichiometric concentrations of enzyme and substrate [4,9].

This is an important prediction of the collision model [12]; in the absence of collision with a second enzyme, tracking will continue unabated up to the end of the DNA [8]. Only when a large excess of enzyme over DNA is present, where the non-specific DNA is loaded with enzyme, can some cleavage be measured [9,13]. Nonetheless, ATP hydrolysis has been measured on single-site substrates [8,10,14] and short DNA oligomers [14], in the absence of DNA cleavage. This is expected, as a one-dimensional process cannot identify distant sites without continuously 'patrolling' the lattice, regardless of any success. However, a stall in tracking cannot be induced by other DNA-binding proteins. When *lac* repressor was bound at an operator site positioned off-centre between two type-I recognition sites, DNA cleavage was still produced by type-I interactions at the intermediate loci. This suggests that the endonucleases can bypass other tightly bound proteins, presumably by driving them off the DNA (Figure 3b), but cannot bypass a converging type-I enzyme.

In contrast, plasmid DNA carrying a single type-I site can be completely cleaved at stoichiometric protein concentrations [10]. Since collision with non-specifically bound enzymes is unlikely in this situation, how does stalling and cleavage occur? The solution is that DNA tracking has an effect on DNA topology [5,12,15]. Any protein that follows the helical DNA path during translocation will either rotate around the DNA or, if the protein is somehow constrained, will turn the DNA around its helical axis [16]. In the DNA-tracking model (Figure 2), the HsdR subunit cannot rotate around the DNA due to its interaction with HsdS and HsdM [6,7]. Therefore, following the helical DNA path during tracking would introduce a decrease in twist into the expanding loop and an increase in twist into the DNA ahead of the protein [9]. On circular DNA, tracking generates a figure-of-eight structure [8,9], consisting of both expanding and contracting loops (Figure 3c). Since these domains are topologically independent, changes in twist would in turn alter the DNA topology [16]. As tracking progresses, the DNA ahead of the complex in the contracting loop would become positively supercoiled (Figure 3c). Eventually, the DNA would become entangled to such an extent that it would pose too large a thermodynamic barrier to further translocation and the enzyme would stall [9]. Correspondingly, DNA cleavage by *Eco*R124I was enhanced on relaxed plasmid substrates [10]; these are closer along the pathway to positively supercoiled DNA (Figure 3c). Tangled DNA loops equivalent to those in Figure 3(c) are also evident in the electron-microscopic data [8]. On topologically unconstrained linear DNA, the enzyme cannot stall as the changes in twist are dissipated by rotation of one DNA strand around the other. Some interpretations of the enzyme-to-enzyme collision model infer that each translocating species contributes one HsdR subunit towards cleavage of one DNA strand. But on circular DNA at least, protein–protein contacts between type-I enzymes are not required to activate DNA hydrolysis. Instead it would appear that simply stopping translocation is adequate, and that a single type-I complex can cleave both DNA strands.

The topological model highlights an important problem for DNA tracking, and indeed any translocation mechanism that forms a DNA loop, e.g. the methyl-directed DNA-mismatch repair proteins. When type-I enzymes first bind DNA, the expanding DNA loop formed is relatively small (probably 10–50 bp). For tracking to proceed over thousands of base pairs, significant levels of negative twist must be introduced into the expanding loop, possibly 360° for every 10 bp travelled [9,16]. Starting from a small DNA loop, it would be impossible to introduce the necessary changes in twist without the coincident topological strain being relaxed; tracking would fail after only 10 or 20 bp. This also applies to tracking on linear DNA [9]. One option is to introduce a nick into the expanding loop prior to starting translocation [9]; changes in twist could then be released by free rotation of one strand around the other at the DNA nick. This tallies with experimental observations that DNA nicking is an early step in the DNA-cleavage pathway. However, recent data from a number of groups have indicated that DNA cleavage and tracking activities can be uncoupled. This invokes a more complicated explanation involving a DNA gyrase-like nicking–closing reaction [5], but evidence of topoisomerase activity has yet to be found. An alternative model is that the protein only contacts one 'side' of the lattice [9], akin to kinesin moving along microtubules. On a perfectly regular lattice, no twist would need to be introduced. However, the helical repeat of DNA is rarely an integral value and some twist would have to be introduced at each step. Allowing the DNA to rotate freely during motion presupposes that the protein releases the DNA, and even transient dissociations would cause DNA slippage due to diffusion.

Most of the experiments described above have been carried out *in vitro*. Nonetheless, the co-operative interactions and cleavage between sites predicted by the collision model [12] have also been observed *in vivo* [15]. At high intracellular concentrations of enzyme, the co-operative interactions break down, suggesting interaction between specifically and non-specifically bound enzymes. Since phage DNA with a single site is also cleaved, the topological model may be vital *in vivo* (phage DNA must circularize prior to integration into the host genome, and any phage that escapes restriction in its linear form could be cut in its circular form by the topological pathway). The significant post-nuclease phase of ATPase activity recorded *in vitro* [8] would appear to be wasteful if invoked *in vivo*. However, it has been suggested that this process may be part of the host defence mechanism: an infected cell would be altruistically removed from the actively growing population by a rapid cellular depletion of ATP [2]. Alternatively, other cellular nucleases may degrade the DNA to prevent any further DNA tracking and ATP hydrolysis [8].

Collision-dependent DNA cleavage by the type-III enzymes

Type-III restriction endonucleases also recognize asymmetric DNA sequences (e.g. CAGCAG for *Eco*P15I), and require Mg^{2+} ions, ATP and *S*-adenosyl methionine to cut DNA [2]. But, unlike type-I enzymes, DNA cleavage only occurs 25–27 bp on the 3′ side of the site. The first evidence for motion on DNA came from the observation that although T3 phage DNA is susceptible to cleavage by *Eco*P15I, T7 phage DNA is completely refractory to cleavage despite carrying 36 *Eco*P15I-recognition sites [17]. This paradox is due to the relative alignment of the asymmetric *Eco*P15I restriction sites. Surprisingly, all 36 sites on T7 are in the same head-to-tail orientation (i.e. the sequence 5′-CAGCAG-3′ is always found on one strand, 5′-CTGCTG-3′ on the other). In contrast, at least some pairs of sites on T3 DNA are in the reciprocal head-to-head and tail-to-tail orientations. By analysing the cleavage of substrates with alternative site alignments, it was established that DNA is only hydrolysed when a pair of type-III sites are present in a head-to-head orientation [17,18]. Furthermore, when a *lac* operator sequence was inserted into the DNA between two such sites, *lac* repressor blocked DNA cleavage (Figure 3b). Only a mechanism that involves one-dimensional transfer of protein(s) along DNA could be suppressed in this way [18]. The current view of type-III translocation is equivalent to the type-I DNA-tracking model shown in Figure 2(a). The Mod subunits bind the recognition site, whereas the Res subunits bind and translocate the adjoining non-specific DNA by hydrolysing ATP [18,19]. The asymmetry of the recognition site is important in predetermining the direction of translocation. When two tracking enzymes collide in the correct head-to-head polarity, DNA cleavage is generated at loci proximal to the recognition sites (Figure 3a). Despite the similarity of the tracking mechanisms, the levels of ATP hydrolysis by the type-I and type-III enzymes are very different; type-I enzymes hydrolyse two orders of magnitude more ATP than the type-III enzymes to move a given distance [18]. This most probably reflects differences in the efficiency of coupling ATP hydrolysis to motion (i.e. more than one ATP is hydrolysed per step taken along the DNA). The coupling may also alter due to changes in load during translocation, e.g. as positive supercoils build up on circular DNA.

Models for motion along DNA

So far, no X-ray crystallographic data on the type-I or type-III enzymes is available and the exact details of DNA motion remain elusive. However, primary amino acid sequence alignments have revealed that both the HsdR and Res subunits contain sequences with significant homology to seven conserved motifs from superfamily II of DNA and RNA helicases [20] (called DEAD-box motifs, after the single amino-acid code characteristic of motif II). The helicases unwind nucleic acid polymers to provide single-strand intermediates involved in a multiplicity of genetic events [21]. They are active as monomers, dimers and even as multimeric rings of subunits which encircle DNA. Estimates of helicase motion suggest they are highly processive, capable of unwinding at least 30 000 bp before dissociation. So could the HsdR and Res subunits utilize an ATP-dependent strand-separation mechanism when tracking DNA? One assay of DNA helicases measures the separation of short DNA duplexes into the corresponding single-strand fragments [21]. Equivalent assays carried out with type-I and type-III endonucleases have failed to demonstrate strand separation. Although the oligomers are resistant to cleavage, they do support ATP hydrolysis. This suggests that some tracking may be occurring. Failure of the helicase assays may simply reflect differences in the 'loading' of the motor on to the DNA. DNA helicases are of two mechanistic classes: those that need a 3′ flanking strand to start the reaction (3′–5′ helicases) and those that need a 5′ flanking strand (5′–3′ helicases). Accordingly, both classes unwind DNA starting at a predefined end. In contrast, type-I and type-III enzymes bind DNA at specific sites and then translocate a region of adjacent non-specific DNA (Figure 2). The stability of the HsdS–DNA and Mod–DNA complexes may prevent strand separation or, because of the length of the oligomers, the HsdR and Res subunits may only contact the extreme DNA ends. The measurable ATPase activity could represent the enzyme slipping off the DNA end while it attempts to start tracking.

The important modular features of the superfamily II helicases have been revealed by the high-resolution crystal structures of the PcrA and Rep helicases. In both structures, the DEAD-box sequences are clustered into a nucleotide-binding pocket flanked by two RecA-like subdomains. There is no apparent homology to the RecA fold in the endonucleases (this may only be revealed by an X-ray crystallographic structure). However, analysis of restriction enzymes mutated in the DEAD-box motifs showed that these sequences are indispensable [15,19,22]. ATP binding is generally not impaired, but little or no ATP hydrolysis is retained and, in concert, DNA cleavage is impaired. This demonstrates a tight correlation between ATP and DNA hydrolysis. A general mechanism for DNA unwinding, supported by convincing kinetic evidence [21], states that a helicase alternates between tight single-strand and duplex DNA-binding states coupled to the binding and hydrolysis of ATP. A conformation switch between two high-affinity lattice-binding states is analogous to stepping models proposed for kinesin (Chapter 6). Two models have been proposed to explain DNA unwinding. In the 'inchworm' model (Figure 4a), a helicase travels along DNA analogous to a zipper opening a zip. The allosteric switch between single- and double-strand binding occurs on one polypeptide, regardless of the subunit composition of the helicase. The alternative 'active rolling' model (Figure 4b) requires that the alternating lattice-binding states are separated on to two separate proteins or two structurally distinct domains. Relating these models to the type-I and type-III enzymes relies on the subunit stoichiometry of the endonuclease complexes. In the absence of a direct measure of motion on DNA by the endonucleases, either DNA-unwinding model could be accommodated. Furthermore, one cannot rule out

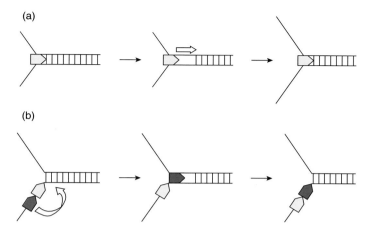

Figure 4. Motion on DNA by strand separation using DNA helicase mechanisms
Single- or double-stranded DNA is represented as black lines, proteins are represented as blue blocks. (a) Inchworm model. The enzyme separates the DNA strands ahead of itself, and then moves forward on to the exposed single-stranded DNA in discrete steps. (b) Rolling model. The enzyme dimer 'walks' along the DNA in discrete steps by concerted alteration of DNA affinity between the two subunits.

the possibility that the HsdS/HsdM or Mod subunits also contribute to the motor mechanisms.

Future prospects

Despite 30 years of effort and a wealth of experimental data, the dynamics of motion generated by the type-I and type-III restriction endonucleases remains poorly characterized. One problem has been that although ATPase and DNA-cleavage activities can be accurately measured, DNA motion has yet to be directly quantified. Furthermore, an ensemble of enzymes undergoing consecutive reactions with equivalent rates, such as DNA tracking, rapidly become unsynchronized. Interpreting steady-state and rapid-quench measurements under these conditions is not trivial. A recent breakthrough has been the analysis of RNA polymerase using single-molecule techniques developed for the classical motor proteins [23]. A DNA template attached to a silica bead is held in the focus of a laser trap. RNA polymerase attached to a glass surface binds the tethered DNA and, in the presence of ribonucleoside triphosphates, translocates the DNA. Movement can be detected optically through movement of the attached bead held in the trap. Measurements of force and velocity have been made that would be difficult to obtain otherwise. The RNA polymerase turns out to be a powerful motor, capable of generating forces up to 25–30 pN, sufficient to overcome the forces opposing transcription *in vivo*. However, one should not ignore the potential of 'classical' experiments. An elegant series of kinetic measurements by Ali and Lohman [24] indicated a physical step size of ≈5 bp for the UvrD helicase, whereas the crystal structure of the closely related Rep enzyme had a DNA 'footprint' of ≈8 bp. This difference is incompatible with a rolling model, where the minimum step size is constrained by the protein footprint on the lattice. Unfortunately, a step size on this scale would probably be swamped by thermal noise in the single-molecule experiments. Key to the future analysis of other DNA-based motors such as the restriction endonucleases will be the application of both techniques. A great deal of work remains to be done, and the next few years promise many exciting developments.

Summary

- *Protein-mediated communications on DNA are universally important. The translocation of DNA driven by a high-energy phosphoryl potential allows long stretches of DNA to be traversed without dissociation.*
- *Type-I and type-III enzymes both use a common DNA-tracking mechanism to move along DNA, dependent on the hydrolysis of ATP.*
- *Type-I enzymes cleave DNA at distant DNA sites (and in some cases close to the site), due to a stall in enzyme motion. This can be due to collision with another translocating type-I enzyme or, on circular DNA, due to an increased topological load. ATP hydrolysis is considerable, and continues after DNA cleavage.*
- *Type-III enzymes only cleave DNA proximal to their sites due to collision between two endonucleases tracking with defined polarity. ATP hydrolysis is less than with the type-I enzymes.*
- *Homology to DNA helicases has been found within the HsdR and Res subunits. Mutagenesis of the DEAD-box motifs affects both ATP hydrolysis and DNA cleavage. This demonstrates a tight link between ATPase and endonuclease activities. A strand-separation mechanism akin to the DNA helicases is a possibility.*
- *The DNA-based motor proteins are mechanistically ill-defined. Further study using some of the techniques pioneered with classical motor proteins will be needed to reveal more detail.*

I thank the Wellcome Trust for financial support.

References

1.   Adzuma, K. & Mizuuchi, K. (1989) Interactions of proteins located at a distance along DNA: mechanism of target immunity in the Mu DNA strand transfer reaction. *Cell* **57**, 41–47
2.   Bickle, T.A. (1993) The ATP-dependent restriction enzymes, in *Nucleases*, 2nd edn. (Linn, S.M., Lloyd, R.S. & Roberts, R.J., eds.), pp. 89–109, Cold Spring Harbor Press, Cold Spring Harbor
3.   Dryden, D.T.F., Cooper, L.P., Thorpe, P.H. & Byron, O. (1997) The *in vitro* assembly of the *Eco*KI type I DNA restriction/modification enzyme and its *in vivo* implications. *Biochemistry* **36**, 1065–1076

4. Rosamund, J., Endlich, B. & Linn, S. (1979) Electron microscopic studies of the mechanism of action of the restriction endonuclease of *Escherichia coli* B. *J. Mol. Biol.* **129**, 619–635

5. Yuan, R., Hamilton, D.L. & Burckhardt, J. (1980) DNA translocation by the restriction enzyme from *Escherichia coli* K. *Cell* **20**, 237–244

6. Kneale, G.G. (1994) A symmetrical model for the domain structure of type I methyltransferases. *J. Mol. Biol.* **243**, 1–5

7. Dryden, D.T.F., Sturrock, S.S. & Winter, M. (1995). Structural modeling of a type I DNA methyltransferase. *Nature Struct. Biol.* **2**, 632–635

8. Endlich, B. & Linn, S. (1985) The DNA restriction endonuclease of *Escherichia coli* B. I: studies of the DNA translocation and the ATPase activities. *J. Biol. Chem.* **260**, 5720–5728

9. Szczelkun, M.D., Dillingham, M.S., Janscak, P., Firman, K. & Halford, S.E. (1996) Repercussions of DNA tracking by the type IC restriction endonuclease *Eco*R124I on linear, circular and catenated substrates. *EMBO J.* **15**, 6335–6347

10. Janscak, P., Adadjieva, A. & Firman, K. (1996) The type I restriction endonuclease *Eco*R124I: overproduction and biochemical properties. *J. Mol. Biol.* **257**, 977–991

11. Szczelkun, M.D., Janscak, P., Firman, K. & Halford, S.E. (1996) Selection of non-specific DNA cleavage sites by the type IC restriction endonuclease *Eco*R124I, *J. Mol. Biol.* **271**, 112–123

12. Studier, F.W. & Bandyopadhyay, P.K. (1988) Model for how type I restriction enzymes select cleavage sites in DNA. *Proc. Natl. Acad. Sci. U.S.A.* **85**, 4677–4681

13. Murray, N.E., Batten, P.L. & Murray, K. (1973) Restriction of bacteriophage λ by *Escherichia coli* K. *J. Mol. Biol.* **81**, 395–407

14. Dreier, J. & Bickle, T.A. (1996) ATPase activity of the type IC restriction modification system EcoR124II. *J. Mol. Biol.* **257**, 960–969

15. Webb, J.L., King, G., Ternet, D., Titheradge, A.J.B. & Murray, N.E. (1996) Restriction by *Eco*KI is enhanced by co-operative interactions between target sequences and is dependent on DEAD box motifs. *EMBO J.* **15**, 2003–2009

16. Liu, L.F. & Wang, J.C. (1987) Supercoiling of the DNA template during transcription. *Proc. Natl. Acad. Sci. U.S.A.* **84**, 7024–7027

17. Meisel, A., Bickle, T.A., Krüger, D.H. & Schroeder, C. (1992) Type III restriction enzymes need two inversely orientated recognition sites for DNA cleavage. *Nature (London)* **355**, 467–469

18. Meisel, A., Mackeldanz, P., Bickle, T.A., Krüger, D.H. & Schroeder, C. (1995) Type III restriction endonucleases translocate DNA in a reaction driven by recognition site-specific ATP hydrolysis. *EMBO J.* **14**, 2958–2966

19. Saha, S., Ahmad, I., Reddy, Y.V.R., Krishnamurthy, V. & Rao, D.N. (1998) Functional analysis of conserved motifs in the type III restriction-modification enzymes. *Biol. Chem.* **379**, 511–518

20. Gorbalenya, A.E. & Koonin, E.V. (1991) Endonuclease (R) subunits of type-I and type-III restriction-modification enzymes contain a helicase-like domain. *FEBS Lett.* **291**, 277–281

21. Lohman, T.M. & Bjornson, K.P. (1996) mechanisms of helicase-catalysed DNA unwinding. *Annu. Rev. Biochem.* **65**, 169–214

22. Davies, G.P., Powell, L.M., Webb, J.L., Cooper, L.P. & Murray, N.E. (1998) *Eco*KI with an amino acid substitution in any one of seven DEAD-box motifs has impaired ATPase and endonuclease activity. *Nucleic Acids Res.* **26**, 4828–4836

23. Wang, M.D., Schnitzer, M.J., Yin, H., Landick, R, Gelles, J. & Block, S.M. (1998) Force and velocity measured for single molecules of RNA polymerase. *Science* **282**, 902–908

24. Ali, J.A. & Lohman, T.M. (1997) Kinetic measurement of the step size of DNA unwinding by the *Escherichia coli* UvrD helicase. *Science* **275**, 377–380

12

Molecular motors in the heart

Fergus D. Davison, Leon G. D'Cruz and William J. McKenna[1]

Department of Cardiological Sciences, St. George's Hospital Medical School, Cranmer Terrace, London SW17 0RE, U.K.

Introduction

The requirement for movement has been of fundamental importance throughout the animal kingdom and although Nature was not able to invent the wheel it did invent the ratchet. This mechanism for producing movement by mechanical contraction at a molecular level is a highly efficient and adaptable process that has been utilized by most animal species. Recent developments in molecular analysis have greatly increased our understanding of the mechanisms involved during the process of muscle contraction. In particular, genetic investigations and molecular modelling of proteins have revealed how the various structural proteins are interacting during molecular contraction. In this essay we describe the contractile system, the muscle proteins involved, and genetic mutations that alter cardiac-muscle function.

The contractile unit

The basic unit that produces contraction in muscle cells (myocytes) is the sarcomere. It consists of a series of sliding, interdigitating structural proteins that are able to shorten when given the appropriate stimulus, normally a neuro-chemical command originating in the brain. A single sarcomere is $\approx 3\ \mu m$ in

[1]*To whom correspondence should be addressed.*

length and up to 5000 may be joined end to end. Large numbers of these sarcomeric concatemers lying together in parallel form a myofibril of which several
are again arranged in parallel to constitute the contractile apparatus of a myocyte, descriptions of which will be found in standard textbooks [1]. The signal
for a muscle to contract normally affects all sarcomeres, causing them to contract in synchrony by approximately 1–2 μm in each case. The total distance of
contraction is a multiple of the number of sarcomeres and can therefore be in
the order of many centimetres. The contractile response can be graded by
neuronal control over the muscle: either by varying the number of myocytes
stimulated to contract or by altering the frequency of motor-nerve firing.

Muscle fibres are divided into two broad types: the slow-contracting
(type-1 or slow-twitch) isoforms are found in tissues where long-term repetitive movements occur and resistance to fatigue is important, e.g. cardiac and
most skeletal muscle. Fast-contracting (type-2) isoforms constitute fast-twitch
fibres, where fast, short-term bursts of activity are needed and the onset of
fatigue is fairly rapid, e.g. extraocular muscles. Both types are found in proportions that reflect specific requirements of the muscle. For example, leg muscles
in animals that are sprinters (e.g. the cat family) contain mostly fast-twitch
fibres, whereas in humans slow-twitch fibres predominate. The fibre types are
distinguished at a molecular level by their myosin ATPase isoenzymes, which
can also be designated as slow or fast. Both types of fibre respire anaerobically
during the first 1–2 min of moderate exercise, as it takes this amount of time
for the cardiopulmonary system to respond. However, slow-twitch fibres are
endowed with a high aerobic oxidative capacity, they have greater numbers of
mitochondria and aerobic respiratory enzymes, a richer capillary supply and
contain myoglobin. In contrast, fast-twitch fibres respire anaerobically by
metabolizing their stored glycogen using the glycolytic pathway. The stage of
development of the organism, and the specific demands the of tissue concerned, determine the proportions of fibres containing these different myosin
isoforms at any particular time. This represents an important adaptive ability
to adjust to changing functional requirements.

Sarcomeric proteins

The structural proteins that comprise the sarcomere are shown in Figure 1.
The thick filaments are composed of 300–400 myosin molecules that make up
the greatest proportion of muscle tissue. The thin filaments are composed of
actin and the troponin–tropomyosin complex, which are critical for regulation
of the contractile response. The myosin light chains are two further molecules
found in lesser amounts; their role is to alter the rate of ATP hydrolysis,
although they are not part of the catalytic site on myosin. The structural
integrity of the sarcomere is maintained by myosin-binding protein that binds
together titin, one of the cytoskeletal proteins, and the myosin heavy chain.

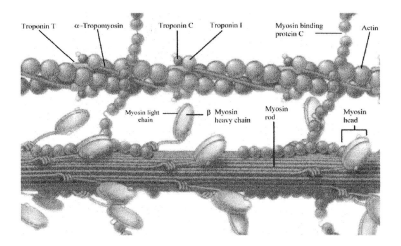

Figure 1. Components of the sarcomere

The lever-arm model of contraction or the crossbridge cycle

The contraction cycle is initiated when the myosin head region forms crossbridges to specific sites on the actin filaments. The positioning of myosin is facilitated by the flexibility of the hinge segment that allows conformational changes to take place. The cycle can be represented as four steps (Figure 2).

(i) ATP bound to myosin is hydrolysed to form a myosin·ADP·P_i complex in the enzymic active site, the myosin head is now in a 'cocked' or high-potential-energy position. This complex has a high level of free energy and a high affinity for actin. Actin has no enzymic activity but it does accelerate the hydrolysis of myosin-bound ATP.

(ii) The myosin·ADP·P_i complex binds to actin and the angle of the myosin head in this crossbridge between the two filaments is 90°.

(iii) After crossbridge formation, ADP and P_i are released and the myosin head then undergoes a conformational change, moving through 45° so that it reaches a final position at 45° with respect to actin. This movement is the power stroke that draws the actin filament past the myosin and causes an overall contraction of the sarcomere when all filaments slide in synchronized movement.

(iv) ATP binds to myosin while it is still crossbridged to actin. This ATP·myosin·actin complex has a much reduced binding affinity, which causes the myosin and actin filaments to dissociate. The cycle repeats as the bound ATP is hydrolysed.

The changes in relative affinity between actin and myosin binding are highly regulated by ATP and ADP+P_i. ATP and ADP+P_i are thus crucial for

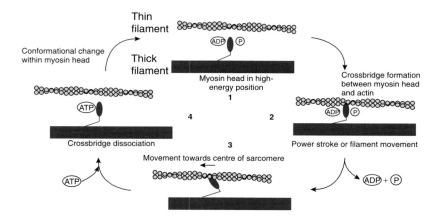

Figure 2. The sarcomeric contraction mechanism

the control of the cycle of crossbridge formation and sliding between filaments. In summary, the presence of ATP brings about dissociation of actin
and myosin. However, once it has been hydrolysed to ADP+P$_i$ within the
enzymic site of myosin, the high binding affinity between myosin and actin is
re-established, enabling the contraction process to continue.

Muscle power output

A single crossbridge cycle moves the actin filament between 13 and 30 nm and
develops a force of about 5×10^{-12} N. When scaled up by several million
crossbridges acting together, remarkable forces can be produced. Recent
research has suggested that the myosin heads may move in a stepping motion
along the actin filament [2]. This study found that several steps were made
from the hydrolysis of a single ATP molecule as the energy from ATP was
released in discreet amounts. Further research is required to determine
precisely how force generation is coupled to the release of ATP. At optimal
contraction lengths muscle cells generate about 3 kg·cm^{-2} or 3×10^5 N·m^{-2}.
The point of optimal force generation occurs when a maximum number of
crossbridges line up between the myosin heads and the actin filaments in each
half of the sarcomere. An increase in strength correlates with a greater cross-
sectional area of muscle fibres. Muscle size is increased by the synthesis of
more myosin, actin and sarcomeres in fast-twitch fibres in response to exercise.
This will increase the number of myofibrils and the diameter of myocytes.
Although this results in larger muscles there is no increase in the number of
myocytes, as this is determined around birth. Muscle contraction accounts for
the majority of the body's ATP consumption, so the process has evolved to be
as efficient as possible. Efficiency is defined by the mechanical work achieved
as a proportion of the chemical energy produced from ATP hydrolysis. In
moderately loaded muscle the maximum efficiency is about 45%, but overall

the average is 20–25% when other muscular reactions, e.g. shivering for heat generation, are considered.

Neuromuscular control

The regulation of muscle movement under neuronal control is known as excitation–contraction coupling. This process involves a signal pathway originating from the action potential of a nerve, transduced at the muscle-cell membrane via a second messenger, which in turn regulates crossbridge cycling. In most cases the control of skeletal muscle ultimately originates within the brain, which is connected to the muscles via motor neurons. A single neuron may divide and connect with up to many thousands of myocytes at neuromuscular junctions, known as end plates. The end plate is a specialized structure that joins the sarcolemma or myocyte cell membrane through which action potentials can be propagated along the myofibril. The sarcolemma interconnects with a sleeve-like arrangement surrounding the myofibril known as the sarcoplasmic reticulum. This network encloses the myofibril, permits communication between sarcomeres via transverse tubules and ensures that action potentials travel within the sarcolemma and spread rapidly throughout the myofibril. The second messenger responsible for precise control of crossbridge cycling is Ca^{2+}. The cytoplasmic Ca^{2+} concentration in resting cells is approximately 10^{-7} M. The action potential depolarizes the sarcolemma via the transverse tubule membranes, which permits an influx of Ca^{2+} from reservoirs in the sarcoplasmic reticulum into the myocyte, increasing the Ca^{2+} concentration to around 10^{-5} M. At this point there is now sufficient Ca^{2+} to interact with the troponin subunits, facilitating the binding of actin and myosin filaments in preparation for the next crossbridge cycle. Ca^{2+} diffusion down the electrochemical gradient is a rapid process taking about 1–2 ms because of the steep concentration gradient and short distances (1 μm) involved. The return of Ca^{2+} into the sarcoplasmic reticulum is achieved by an ATP-driven Ca^{2+} pump.

Cardiac muscle

Cardiac muscle, although structurally and functionally similar to skeletal muscle, has some special features. First, the sarcoplasmic reticulum is much reduced in the heart as the larger diameter of cardiac fibres necessitate relatively high concentrations of stored cytosolic Ca^{2+}, independent of extracellular supply. Secondly, cardiac muscle does not require external stimulation from motor neurons in order to contract; in fact the heart can continue to beat after it is removed from the body. This ability to initiate its own contraction and rhythm is due to specialized groups of cells called pacemakers. The most important of these is the sinoatrial node situated in the right atrium. Finally, myocardial cells are short, branched, interconnected with adjacent cells and connected by electrical synapses or gap junctions. This feature enables a large mass of cardiac myocytes to contract autonomously as single cell block.

Myosin

The myosin filament has both structural and enzymic properties and consists of two identical intertwined chains. Within the head region there is a catalytic site where the chemical energy of ATP is released to provide the force needed during sarcomeric shortening. The long tail regions are α-helices forming a coiled structure, which in turn interacts with the tail regions of other myosin molecules. The myosin molecule is composed of six polypeptide chains: two heavy chains and four light chains, where each heavy chain is bound to two light chains in the head region of the molecule. Myosin can be expressed in the muscle in several different isoforms, which are encoded by separate genes, the combination of isoforms being specified by the requirements of the muscle. The predominant isoform found in humans is β-myosin, associated with type-2 muscle (see below), i.e. skeletal and cardiac muscle. α-Myosin is found in the embryonic heart and at low levels in skeletal muscle, whereas the other major adult isoforms, 2a, 2b and 2x, are all associated with type-1 muscle. The proportion of these isoforms may change, particularly during development, and is under hormonal control, especially within heart muscle, e.g. by thyroxine, atrial natriuretic factor and angiotensin.

The troponin–tropomyosin complex

For the heart to function as a pump, a mechanism is required that can cyclically switch on and off crossbridge formation in each sarcomere and modulate overall cardiac performance. In skeletal and cardiac muscle the control of initiation, inhibition and speed of contraction is undertaken by the troponin–tropomyosin complex. Troponin is a complex of three closely associated proteins, i.e. troponin T (Tn T, tropomyosin binding), troponin C (Tn C, calcium binding) and troponin I (Tn I, inhibition of tropomyosin binding), and the complex is distributed along the length of the thin filament (Figure 3). During diastole (low calcium) Tn I binds to actin and the troponin trimer 'pulls' on the tropomyosin subunit, displacing it from its original location to occupy the myosin-binding site on actin. This prevents formation of crossbridges between actin and myosin and hence the muscle is relaxed. When the calcium concentration increases during systole, Tn I is released from actin and the tropomyosin filament then moves from the myosin-binding site, facilitating the formation of the actomyosin crossbridge and initiating muscle contraction.

The various isoforms of the troponin–tropomyosin complex are muscle-fibre specific, e.g. the cardiac isoform of T T is not found in skeletal muscle. They belong to separate multigene families (that vary slightly in their coding sequence), which are discussed in the following section with emphasis on cardiac isoforms.

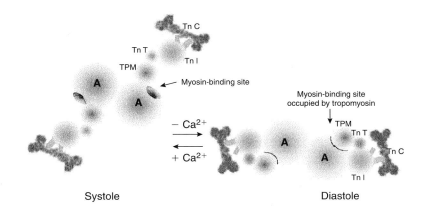

Systole   Diastole

Figure 3. The regulation of contraction and relaxation in muscle
Cross-sectional schematic view of the sarcomere showing the spatial relationships of actin (A), Tn T, tropomyosin (TPM), Tn C and Tn I. The cleft where myosin binds to actin is also shown. A drop in Ca^{2+} concentration causes a conformational change in Tn C, allowing Tn I to bind to actin, which pulls on TPM. In the process myosin is allosterically inhibited from binding to actin.

Tn T

The Tn T subunit is a rod-like molecule whose principal function is to bind Tn I at one site and tropomyosin at two sites. The gene encoding cardiac isoform (*cTn*) has 16 exons. Its primary transcripts are mRNAs encoding four isoforms; cTn T1, cTn T2, cTn T3 and cTn T4. These mRNAs arise from the inclusion or exclusion of exons 4 and 5 (Table 1). The default pathway is to exclude exon 5, which accounts for the predominance of cTn T3 as the major isoform found in adult heart.

Tn C

The cardiac and slow skeletal isoforms of Tn C are encoded by the same gene in skeletal and cardiac muscle. cTn C, an 18.4 kDa protein, is a member of the Tn C superfamily and is evolutionarily related to a number of calcium-binding proteins (including calmodulin, calbindin, myosin light chain and parvalbumin). Tn C adopts a dumb-bell shape with two globular Ca^{2+}-binding

Table 1. Isoforms of human cTn T

cTn T isoform	Sequence composition	Tissue expression
cTn T1	All 16 exons	Fetal heart
cTn T2	Exon 4 omitted	Fetal heart (low levels)
cTn T3	Exon 5 omitted	Adult heart (predominant form)
cTn T4	Exons 4 and 5 both omitted	Fetal heart but also re-expressed in the adult failing heart

Alternative RNA splicing involving exons 4 and 5 accounts for the heterogeneity in expression.

sites on opposite ends linked by a 31-residue central α-helix (Figure 4). Sites III and IV on the C-terminus are similar in the fast- and slow-twitch isoforms; these high-affinity binding sites are always occupied with either Ca^{2+} or Mg^{2+} under physiological conditions. However, the lower-affinity N-terminus shows marked variation between the slow- and fast-twitch isoforms because only the fast-twitch isoform shows occupancy of sites I and II, whereas the cardiac and slow-twitch isoforms show occupancy of site II only. Further studies in which the site-II calcium-binding residues in cTn C were genetically engineered into site I while swapping site-I residues for site II showed complete inactivation of the cTn C molecule [3,4]. These results imply that the reduced affinity for Ca^{2+} at the N-terminal end of Tn C appears to be critical for the regulatory function in heart muscle. The inability of site I in cTn C to co-ordinate with Ca^{2+} is due to the substitution of a negatively charged aspartic acid residue (Asp-29) in fast skeletal Tn C by a leucine residue in the cardiac/slow skeletal isoform, and to an insertion of a valine residue (Val-28) in cTn C (Figure 4) [5].

In fast skeletal Tn C, the binding of Ca^{2+} to sites I and II results in a conformational change whereby the helices in the N-terminal region move away from each other. This exposes a short sequence of hydrophobic residues (Figure 4) that are normally buried within the tertiary structure of the protein. The exposure of the hydrophobic region alters the interaction between skeletal Tn I and fast skeletal Tn C, disrupting the interaction between actin and Tn I. In the process, tropomyosin moves away from the myosin-binding site allowing interaction of myosin and actin, permitting muscle contraction.

However, the binding of Ca^{2+} to site II of cTn C does not result in the 'opening' of the structure to the same extent as observed in the fast skeletal isoform of Tn C [6]. In the presence of Ca^{2+}, there is a 4-fold greater affinity between fast skeletal Tn C and skeletal Tn I, compared with the affinity between cTn C and Tn I [7].

Increasing the sensitivity of cTn C to Ca^{2+} results in the symptomatic relief of congestive heart failure. Levosimendan, an experimental drug, has been shown to interact directly with Asp-88 of cTn C while occupying a space surrounded by Met-81, Met-85 and Phe-77 of cTn C. These residues are situated around the hydrophobic region shown in Figure 4. Knowledge of the protein–ligand spatial relationships and molecular sequence data will help in the understanding of the precise conformation of the binding pockets within molecules and hence aid in the design of more effective drugs with enhanced binding capacities.

Tn I

The Tn I gene family comprises three isoforms, one cardiac and two skeletal, the latter designated as Tn I-fast and Tn I-slow with a high degree of homology existing between them. Although cTn I predominates in adult heart, the slow isoform is expressed in heart tissue early in development [8–10]. cTn I is

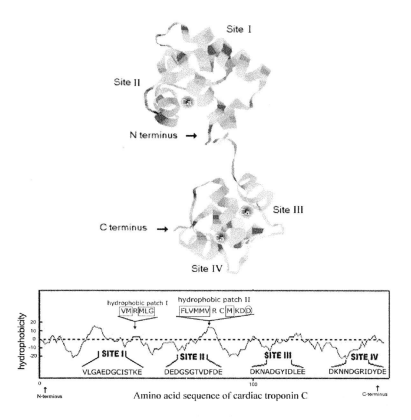

Figure 4. The dumb-bell-shaped cTn C molecule
This isoform only binds Ca^{2+} at sites II, III and IV and not at site I due to a substitution of negatively charged residues with valine and leucine. High Ca^{2+} concentrations cause a conformational change in Tn C, exposing the normally buried hydrophobic amino acid residues (boxed), promoting interaction with Tn I and thereby releasing the bond between Tn I and actin. The circled aspartic acid (D) interacts directly with the drug Levosimendan (see text).

crucial for normal functioning of the heart muscle and in cTn I knockout mice the slow skeletal isoform of Tn I compensates for the absence of the cardiac isoform during fetal development. However, 15 days after birth, expression of slow skeletal isoforms declines rapidly and the mice die by day 18 [11].

Human cTn I is a 23.8 kDa globular molecule which binds to Tn C, Tn T and actin, inhibiting muscle contraction in the absence of Ca^{2+}. In the presence of Ca^{2+}, conformational changes in Tn C are transmitted to Tn I to release its inhibitory binding to actin. There are subtle functional differences between the skeletal and cardiac isoforms of Tn I. Skeletal Tn I strongly inhibits actomyosin ATPase in the absence of tropomyosin, whereas the inhibitory effect of cTn I is much weaker. The integrity of the Tn I–Tn C complex is dependent on Ca^{2+}, suggesting that the Tn C binding or interaction sites on Tn I play a role in the transmission of signal from Tn C to the rest of the thin filament.

Tropomyosin

Tropomyosin is a rod-like, α-β helical heterodimer spanning seven actin residues. In the relaxed (low-Ca^{2+}) state, tropomyosin occupies the myosin-binding sites on the actin residues preventing the interaction between actin and myosin (Figure 3). There are four known tropomyosin genes in humans, designated *TPM1*, *TPM2*, *TPM3* and *TPM4*. These four genes code for different isoforms that are expressed in a tissue-specific manner and regulated by alternative splicing mechanisms. The vertebrate α-tropomyosin gene consists of 15 exons, 10 of which are subject to alternative splicing. In addition, some tropomyosin isoforms arise due to transcription from alternative promoters.

The muscle-specific isoforms of the tropomyosin genes are differentially expressed in various muscle-fibre types and in cardiac tissue. TPM1 is predominantly expressed in fast-twitch fibres and the heart. On the other hand, the non-muscle isoform of α-tropomyosin, TPM3, has a higher level of expression in slow-twitch muscle fibres. TPM2 is expressed mainly in fibroblasts and epithelial muscle cells.

Gene mutations in sarcomeric proteins

Inherited mutations in genes encoding sarcomeric proteins have been well described as the defects that they produce in muscle can result in serious disease [12]. This is particularly true in the case of heart muscle, where extraordinary demands are made upon this organ during the lifetime of an individual. Mutations in genes encoding sarcomeric proteins lead to defective heart-muscle proteins and are collectively the main cause of hypertrophic cardiomyopathy (HCM). The is an autosomal dominant disease affecting about 1 in 500 of the population and is one of the most common causes of sudden death in athletes and individuals younger than 25 years. The same genetic mutations in β-myosin found in the heart are also likely to occur in the skeletal muscles of individuals with HCM, but any serious skeletal muscular defect is rare. This is because skeletal muscle bears a much lesser burden in terms of contraction frequency than the heart. Cardiac-muscle hypertrophy develops in response to mechanical stress, stimulating stretch receptors in the cardiac wall, and is essentially an adaptive or compensatory process by the heart. This stress is induced by contractile dysfunction within sarcomeres containing the defective protein and is a gradual process taking many years. Mechanical stress imposed on the heart induces the expression of a wide range of cytokines and growth factors, e.g. transforming growth factor β, insulin-like growth factor 1 and interleukin 1B, but the complex interplay between these factors during hypertrophy is not fully understood [13]. The disease is characterized by increased left-ventricular-wall thickness, predominantly in the intraventricular septum, and may be associated with right-ventricular hypertrophy. Characteristic histological changes include myocyte and myofibrillar disarray with increased loose connective tissue and fibrosis. These features enable familial HCM to be distinguished from cardiac hypertrophy due to normal physiological factors. Over 100 different disease-causing mutations have been found within the sarcomeric-protein genes, indicating considerable genetic heterogeneity in HCM [14]. The mutations identified have been predominantly missense mutations but also include deletions or mutations leading to mRNA missplicing. As a result, there are changes to the amino acid sequence of the respective sarcomeric proteins that compromise actin–myosin interaction. The phenotype of the disease can vary markedly, and mutational analyses of sarcomeric genes in conjunction with clinical assessments have shown that certain mutations indicate a more serious prognosis in HCM. However, many mutations are not lethal and affected individuals may or may not have a normal life expectancy depending on the extent to which a specific mutation interferes with structure and function [15].

Myosin mutations

The β-myosin heavy chain constitutes the bulk of cardiac myosin and mutations in the gene encoding this protein are estimated to account for around 30% of all HCM mutations. Its gene has 40 exons, but only the first 23 are affected by mutations with deleterious consequences. Within these 23 exons are the exons that encode the major active sites of the myosin molecule. These exons contain 'hot spots' for mutations and over 50 different mutations have been identified so far. These hot spots fall within four regions critical for the normal function of the head region of the molecule: (i) the ATP-binding site, (ii) the actin-binding sites, (iii) the interface between myosin heavy and light chains and (iv) the hinge region of the myosin head where motive force is produced. Mutations in these locations are associated with a more malignant phenotype and premature sudden death. Conversely, other mutations are more benign and may have little effect on patient survival. Mutations that introduce a new amino acid with altered charge or hydrophobicity are typically associated with a poorer prognosis. Further changes to myosin function can be expected on the basis of modifications in the size and shape of the altered amino acid, e.g. the presence of bulky side chains or aromatic rings. The overall penetrance of the disease phenotype will be further modified by epigenetic events, including the effect of the normal allele, physiology and environmental factors, e.g. levels of physical exertion.

Troponin mutations

Approximately 15% of patients with HCM were found to carry mutations in the *cTn T* gene. These patients typically exhibit a milder phenotype, but there is an increased incidence of sudden death. A comparison of the amino acid sequences of the regions coded by exons 8–16 of *cTn T* shows little divergence between species, implying that this region has a functional significance critical to the contractile function of heart muscle. This region also contains important binding sites between cTn T and tropomyosin. Therefore, it is not surprising that mutations in HCM patients have been shown to lie between exons 8 and 16 of the *cTn T* gene. There are no crystallographic data to provide a structure of the cTn T protein, hence it is difficult to ascertain the significance of mutations with respect to the interaction of cTn T with other subunits of the troponin–tropomyosin complex. However, some mutant *cTn T* genotypes (Ile-79→Asn, Arg-92→Glu or ΔGly160) found in cases of HCM were tested by transfecting constructs into quail myotubes. It was found in these transfectants that the physiological concentrations of Ca^{2+} were insufficient for contractile function [16]. In addition, the two missense mutations, Ile-79→ Asn and Arg-92→Glu, independently doubled the unloaded shortening velocity, suggesting that the onset of HCM could be due to increased energy demands on the heart. Possibly cTn T is able to alter the rate of crossbridge detachment and therefore has a greater role in the modulation of the contractile performance than considered previously.

There are two important reactive sites on cTn I, one of which interacts with Tn C and an inhibitory site that binds to actin. *cTn I* mutations have been located in patients with HCM and one of these mutations occurs within the inhibitory domain, resulting in functional changes to the molecule. Further evidence comes from site-directed-mutagenesis studies on the inhibitory domain (implicated in the binding of actin and Tn C) of rabbit skeletal Tn I, which showed that single point mutations were sufficient to abolish activity of the Tn I molecule [17].

Recently, *cTn C* missense mutations were found in idiopathic dilated cardiomyopathy [18]. These missense mutations occurred within site II, involving the Ca^{2+}-binding site of the Tn C molecule.

Tropomyosin mutations

Mutations in the *TPM1* gene are also implicated in HCM and account for ≈3% of HCM cases. Like those in cTn T, they are characterized by relatively mild hypertrophy and an increased incidence of sudden death. The tropomyosins are not only important as components of muscle, but have been implicated in oncogenesis. *TPM1* is altered in a number of human breast and prostate cancer cell lines and recent data suggest that it may also act as a tumour-suppressor protein [19,20].

Perspectives

The molecular machinery responsible for producing movement in the sarcomere is susceptible to a large number of different mutations in the proteins involved in its assembly. The importance of screening individuals for mutations within genes coding for these sarcomeric components cannot be underestimated when familial cardiomyopathy is suspected. Their identification enables screening of other family members who may not yet exhibit any clinical abnormalities but can still carry the mutated gene (hence the term incomplete penetrance of the diseased phenotype). They can then be assessed by clinical criteria, e.g. electrocardiography and echocardiography. Such individuals may develop symptoms later in life, but since HCM carries an associated risk of sudden death, screening for mutations is of significant clinical benefit in terms of anticipating preventative medical care and genetic counselling.

Summary

- *The contractile apparatus of muscle is a highly efficient and adaptable mechanism for producing movement and is exploited throughout the animal kingdom.*
- *Molecular biology is yielding important insights into the intricate functioning of muscular contraction, especially in the heart, and in explaining the genesis of inherited myopathies.*
- *Mutational analyses of the sarcomeric-protein genes in conjunction with clinical assessment have shown that certain mutations indicate a more serious prognosis in HCM. Detecting these mutations in individuals is important for screening other family members for the disease.*
- *Understanding the contractile apparatus at the molecular level could contribute to the design of more effective drugs for treatment of cardiac diseases.*

Further reading

Aitken, A. (1990) *Identification of Protein Consensus Sequences: Active Site Motifs, Phosphorylation, and Other Post-Translational Modifications*, Ellis Horwood, Chichester

Haber, E. (1995) *Molecular Cardiovascular Medicine*, Scientific American, New York

References

1. Berne, R.M. & Levy, M.N. (eds.) (1996) *Principles of Physiology*, 2nd edn., Mosby, St Louis
2. Kitamura, K., Tokunaga, M., Iwane, A.H. & Yanagida, T.R. (1999) A single myosin head moves along an actin filament with regular steps of 5.3 nanometres. *Nature (London)* **397**, 129–134
3. Sweeney, H.L., Brito, R.M.M., Rosevear, P.R. & Putkey, J.A. (1990) The low-affinity Ca^{2+}-binding sites in cardiac slow skeletal-muscle troponin-C perform distinct functions - site-1 alone cannot trigger contraction. *Proc. Natl. Acad. Sci. U.S.A.* **87**, 9538–9542
4. Gulati, J., Babu, A. & Su, H. (1992) Functional delineation of the Ca^{2+}-deficient EF-hand in cardiac-muscle, with genetically engineered cardiac-skeletal chimeric troponin-C. *J. Biol. Chem.* **267**, 25073–25077
5. Van Eerd, J. & Takahashi, K. (1975) The amino acid sequence of bovine cardiac tamponin-C. Comparison with rabbit skeletal troponin-C. *Biochem. Biophys. Res. Commun.* **64**, 122–127
6. Spyracopoulos, L., Li, M.X., Sia, S.K., Gagne, S.M., Chandra, M., Solaro, R.J. & Sykes, B.D. (1997) Calcium-induced structural transition in the regulatory domain of human cardiac troponin C. *Biochemistry* **36**, 12138–12146

7.   Liao, R.L., Wang, C.K. & Cheung, H.C. (1994) Coupling of calcium to the interaction of troponin-I with troponin-C from cardiac muscle. *Biochemistry* **33**, 12729–12734

8.   Sasse, S., Brand, N.J., Kyprianou, P., Dhoot, G.K., Wade, R., Aria, M., Periasamy, Yacoub, M.H. & Barton, P.J. (1993) Troponin-I gene-expression during human cardiac development and in end-stage heart failure. *Circ. Res.* **72**, 932–938

9.   Hunkeler, N.M., Kullman, J. & Murphy, A.M. (1991) Troponin-I isoform expression in human heart. *Circ. Res.* **69**, 409–414

10.   Bhavsar, P.K., Dhoot, G.K., Cumming, D.V.E., Butler-Browne, G.S., Yacoub, M.H. & Barton, P.J. (1991) Developmental expression of troponin-I isoforms in fetal human heart. *FEBS Lett.* **292**, 5–8

11.   Huang, X.P., Pi, Y.Q., Lee, K.J., Henkel, A.S., Gregg, R.G., Powers, P.A. & Walker, J.W. (1999) Cardiac troponin I gene knockout - a mouse model of myocardial troponin I deficiency. *Circ. Res* **84**, 1–8

12.   Schwartz, K., Carrier, L., Guicheney, P. & Komajda, M.R. (1995) Molecular-basis of familial cardiomyopathies. *Circulation* **91**, 532–540

13.   Schaub, M.C., Hefti, M.A., Harder, B.A. & Eppenberger, H.M.R. (1997) Various hypertrophic stimuli induce distinct phenotypes in cardiomyocytes. *J. Mol. Med.* **75**, 901–920

14.   Marian, A.J. & Roberts, R.R. (1995) Recent advances in the molecular-genetics of hypertrophic cardiomyopathy. *Circulation* **92**, 1336–1347

15.   Spirito, P., Seidman, C.E., McKenna, W.J. & Maron, B.J. (1997) Medical progress - the management of hypertrophic cardiomyopathy. *N. Engl. J. Med.* **336**, 775–785

16.   Sweeney, H.L., Feng, H.S.S., Yang, Z.H. & Watkins, H. (1998) Functional analyses of troponin T mutations that cause hypertrophic cardiomyopathy: insights into disease pathogenesis and troponin function. *Proc. Natl. Acad. Sci. U.S.A.* **95**, 14406–14410

17.   Strauss, J.D., Van Eyk, J.E., Barth, Z., Wiesner, R.J., Ruegg, J.C. & Kluwe, L. (1996) Recombinant troponin I substitution and calcium responsiveness in skinned cardiac muscle. *Pflugers Arch. Eur. J. Physiol.* **431**, 853–862

18.   Liao, R., Gwathmey, J.K. & Wang, C.K. (1998) Functional significance of two missense mutations in troponin C found in human myocardium with idiopathic dilated cardiomyopathy. *Circulation* **98**, 3288 (abstract)

19.   Franzen, B., Linder, S., Uryu, K., Alaiya, A.A., Hirano, T., Kato, H. & Auer, G.(1996) Expression of tropomyosin isoforms in benign and malignant human breast lesions. *Br. J. Cancer* **73**, 909–913

20.   Braverman, R.H., Cooper, H.L., Lee, H.S. & Prasad, G.L. (1996) Anti-oncogenic effects of tropomyosin - isoform specificity and importance of protein-coding sequences. *Oncogene* **13**, 537–545

13

The roles of unconventional myosins in hearing and deafness

Richard T. Libby and Karen P. Steel[1]

*MRC Institute of Hearing Research, University Park, Nottingham
NG7 2RD, U.K.*

Introduction

Unconventional myosins are crucial for the proper development and function
of several sensory systems (for a detailed discussion of unconventional
myosins see Chapter 4 in this volume by Kalhammer & Bähler). Specifically,
mutations in unconventional myosins have been found to cause deafness and
retinal degeneration in humans, and similar pathologies can be found in their
animal models. Here, we will concentrate our discussion of myosins on their
roles in hearing and balance; in particular, we will examine the function of
unconventional myosins in the cells in the inner ear that are responsible for
transducing sound and vestibular information into neural code, the hair cells.

There are reasons for studying the role of unconventional myosins in the
inner ear beyond the fact that mutations in them can cause human disease. The
inner ear is a good model system for assessing the function *in vivo* of unconventional myosins for several reasons. First, at least five unconventional
myosins are expressed in the inner ear, each having a unique expression pattern. Second, the anatomy and basic physiology of the inner ear are well
understood, and there are well-defined techniques available to ascertain

[1]*To whom correspondence should be addressed.*

whether the inner ear is functioning normally; thus, there are established experimental paradigms available to analyse the effects of disrupted myosin function. Finally, there are already several mouse models where a mutation in an unconventional myosin is known to affect the normal function of the inner ear. These mouse models can be, and have been, used to ascertain the function of these molecules *in vivo*.

Hearing and balance

The inner ear contains the apparatus that is responsible for hearing and balance. These two systems are physically separated in the inner ear into distinct areas. In the mammalian inner ear, sound waves are transduced by specialized hair cells in the cochlea and balance information is transduced by a similar type of specialized cell in the vestibular system (which is subdivided into five distinct areas). Although these two systems transduce very different stimuli and are physically separated, they both rely on the same specialized cell, the hair cell. An overly simplistic view of the anatomy of the sensory epithelium of the inner ear is that it consists of two basic cell types, hair cells and the supporting cells. The supporting cells function, as their name suggests, to support the hair cell. Their main function is to provide the proper structural, physiological and metabolic environment for hair cells. The hair cell contains the molecular components necessary for transducing sound and vestibular information (Figure 1a). The basal region of the hair cell contains the synaptic machinery necessary to synapse with neurons of the VIII cranial nerve. The hair cell's apical surface is even more specialized than the basal surface; protruding from the apical surface is the structure that detects the stimulus, the hair bundle (Figure 1b). This consists of a highly ordered alignment of stereocilia (stereocilia are filled with actin and are not true cilia), the exact arrangement depending on the species and whether the hair cell is in the cochlea or in the vestibular system. Hair bundles are a highly ordered, stepped arrangement of stereocilia. Individual stereocilia are cross-linked to each other by extracellular extensions called side links. There is a single specialized connection (tip link) at the tip of the stereocilia that is thought to be important in transduction. The stereocilia are anchored in an actin-rich structure that lies just below the apical surface of the hair cell, the cuticular plate. The proper organization and maintenance of the hair bundle is required for normal hearing and balance. As we will see, unconventional myosins play several critical roles in the development and function of the inner ear and, more specifically, the hair cell and its stereociliary bundle.

Myosin VIIa

Human myosin VIIa is predicted to be 2215 amino acids in length and its gene (*MYO7A*) contains 48 coding exons [1,2]. The protein contains a terminal head domain, four IQ motifs, a coiled-coil domain and three myosin tail homology

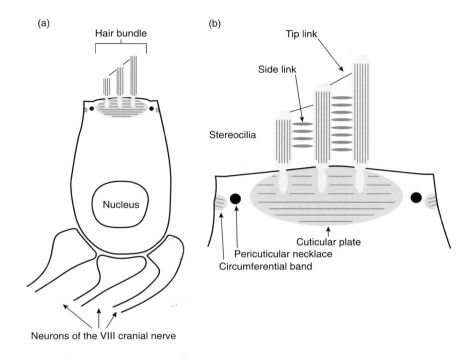

Figure 1. Anatomy of the mammalian hair cell
(a) Entire hair cell. (b) An enlargement of the apical region; note that there are three actin-rich compartments, stereocilia, cuticular plate and circumferential band (actin is represented by dark blue lines).

4 (MyTH4) domains. Myosin VIIa is unique in that it is the only unconventional myosin to contain two talin homology domains (talin is a member of the band-4.1 protein superfamily). Talin-like domains are thought to interact with membranes. As with other members of the superfamily that contain a coiled-coil domain, myosin VIIa is thought to form a homodimer [3]. In mammals the protein is expressed in many different adult and developing tissues and appears to be a common component of cilia and microvilli [4]. Mutations in *MYO7A* can cause either syndromic deafness (deafness linked with other non-inner-ear pathologies) or non-syndromic deafness (pathology confined to the ear). As we will see, different mutations in *MYO7A* can be correlated with the level of hearing impairment, i.e. some lead to progressive deafness and some lead to a profound congenital deafness. A careful analysis of the mutations causing disorders in humans and mice can help us to better understand the function of myosin VIIa. However, one must be careful when correlating genotype with phenotype because the patient's genetic background can affect the severity of the phenotype caused by a particular mutation.

Non-syndromic deafness

Mutations in *MYO7A* cause both dominant and recessive forms of non-syndromic deafness. Analysis of the recessive cases of non-syndromic deafness clearly show that one normal allele is sufficient to enable normal function, i.e. 50% or less of the normal protein level of myosin VIIa is enough to allow normal hair-cell physiology [5]. For a mutation in *MYO7A* to be dominant, the mutant allele must affect the function of the product of the normal allele, and when this has a deleterious effect on myosin VIIa function it is called a dominant negative mutation. Thus far, only one mutation has been identified in *MYO7A* that causes dominant non-syndromic deafness; *DFNA11* [6] (note that non-syndromic deafness loci are given standard names; DFN followed by either A for dominant or B for recessive and then a number). This mutation is an in-frame deletion resulting in loss of three residues within the coiled-coil region thought to be responsible for myosin VIIa homodimer formation. Presumably, the lack of these three residues does not prevent the product of the mutant allele from dimerizing with the product of the normal allele, but the mutation may effect the tertiary structure of the protein, rendering the entire dimer functionless (a dominant negative).

To date, four mutations have been identified in *MYO7A* that lead to recessive non-syndromic deafness (*DFNB2*) [3,7]. One of these is particularly interesting because it demonstrates how the analysis of mutations can provide insight into the roles of certain domains within proteins but that it is not always clear why a mutation causes a disease. In one family, a missense mutation (where one amino acid is substituted for another) in the last nucleotide of an exon was shown to cause profound deafness [3]. Mutations in the last nucleotide of an exon are known to affect splicing and in other genes similar mutations can result in the whole exon being spliced out of the transcript. Exon 15 of *MYO7A* (the exon that would be affected) encodes the actin-binding site, whose absence would severely affect the protein's function. However, the reason this mutation causes deafness is not completely clear since 40% or more of the transcripts appear to be full length and presumably encode enough myosin VIIa for normal function. The amino acid change, a methionine to an isoleucine, is not believed to affect the function of the protein, despite being close to the actin-binding site. The authors give two reasons for this [3]. First, methionine and isoleucine have similar properties and this residue is not conserved across the myosin family, suggesting that it is not a critical residue. Secondly, analysis of the three-dimensional structure of the actin-binding site in muscle myosin predicts that this residue is not important in actin binding. So how do we explain why individuals homozygous for this mutation are deaf? It may simply be that the mutation causes the exon to be spliced out in the vast majority of transcripts and/or that the prediction that this amino acid is not crucial for myosin VIIa to function is incorrect. However, it is still possible to explain the deafness even if the analysis of the mutation is correct, i.e. that 40% of *MYO7A* transcripts are correctly spliced and 60% are missing exon 15. The incorrectly spliced *MYO7A* transcript results in a 46-amino acid in-frame deletion that probably affects the tertiary structure of myosin VIIa. A disruption of the tertiary structure could alter the coiled-coil domain just 220 amino acids downstream from the actin-binding site [3]. As with the dominant mutation discussed above, this abnormal myosin VIIa could be acting in a dominant negative fashion by dimerizing with normal myosin VIIa. Thus for this mutation, it is still unclear what its exact effects are on myosin VIIa function. Furthermore, the analysis of this mutation points out that we still have a lot of questions remaining about all levels of the biology of myosin VIIa.

Syndromic deafness

Usher syndrome type 1B (USH1B)

Usher syndrome is one of the most common forms of syndromic deafness and is characterized by hearing loss, retinitis pigmentosa and, in some types, vestibular dysfunction. The most severe form of the disease is type 1; USH1 patients have profound congenital deafness, vestibular dysfunction and the onset of retinitis pigmentosa prior to puberty. The disease types are further divided (e.g. USH1A and USH1B) based on genetic-mapping studies which have shown that several genetic loci cause the same type of Usher syndrome. Mutations in *MYO7A* are responsible for one subtype, USH1B [1].

There are at least 35 different mutations in *MYO7A* that lead to USH1B. The effects of these mutations are spread over the entire length of myosin VIIa, with the majority of them concentrated in the protein's head region (Table 1). Nonsense, missense, deletion and splice-site mutations have all been identified. Missense mutations that lead to USH1B can be used to help elucidate the importance of particular residues within myosin VIIa. One example is a base substitution (G→A) that causes a glycine residue (2137) to be changed to glutamic acid (a non-conservative substitution). This area of myosin VIIa shows homology to the membrane-binding domain of members of the band-4.1 protein superfamily and, therefore, is thought to be involved in myosin VIIa interactions with membranes. Thus this mutation may significantly affect the function of myosin VIIa by preventing it from being properly targeted within the cell. Furthermore, it shows that this residue is critical for proper function of myosin VIIa.

Atypical Usher syndrome

Patients whose clinical features do not allow them to be classified as having one of the three main forms of Usher syndrome are classified as having atypical Usher syndrome. In two siblings diagnosed with atypical Usher syndrome, the underlying cause was found to be a mutation in *MYO7A* [8]. Unlike the profound congenital deafness and the early onset of retinitis pigmentosa associated with USH1B, a progressive hearing loss began during infancy and retinitis pigmentosa was not diagnosed until after puberty. Both

Table 1. Known mutations in *MYO7A*

Disease	Type of mutation	Location in myosin VIIa	
		Head	**Tail**
DFNA11	Deletion		2658–2666del(IF) (CC)*
DFNB2	Insertion		Val1199insT (FS 28 aa)
	Missense	Arg244→Pro	
		Met599→Ile*†	
	Intronic	IVS3−2a→g	
USH1B	Deletion	75delG(FS 4 aa)	724delC (IQ)
		120delG(FS)	2065delC (Talin)
		532delA(FS 14 aa)	2119–2215del (Talin)
		652–657del(IF)	
		809delC(FS)*	
	Insertion	468+Gln(3 bp ins.)	
	Missense	Gly25→Arg	Ala826→Thr (IQ)
		Thr165→Met	Gly955→Ser
		Gly214→Gln	Arg1240→Gln
		Arg212→His	**Arg1602→Gln (Talin)***
		Arg212→Cys	Arg1743→Trp
		Arg302→His	Gly2137→Glu (Talin)*
		Ala397→Asp	
		Gly450→Glu	
		Pro503→Leu	
		Arg756→Tyr	
	Nonsense	Cys31→Stop	Arg1861→Stop
		Arg150→Stop	
		Gln234→Stop	
		Glu314→Stop	
		Tyr333→Stop	
		Cys628→Stop	
		Arg634→Stop	
		Arg669→Stop	
		Ile668→Stop	
	Intronic	IVS5+1g→a	IVS27−1g→c
		IVS13−8c→g	IVS29−2t→a
		IVS18+1g→a	
Atypical USH	Missense	Leu651→Pro	**Arg1602→Gln (Talin)***

*These amino acid changes are discussed in the text.

†Mutation may also result in exon 15 being spliced out of the transcript.

There are 35 different mutations in *MYO7A* that lead to deafness. These deafnesses can be syndromic (USH1B and atypical Usher syndrome) or non-syndromic (DFNA11 and DFNB2). The domains affected by mutations in the tail are noted in parentheses following the mutation information. The substitution shown in bold is associated with two clinically distinct forms of (contd.) ☞

siblings are compound heterozygotes for mutations in *MYO7A*. The mutation in one allele leads to a non-conservative leucine-to-proline substitution at residue 651. Position 651 is a highly conserved residue within the motor domain and is thought to be important in the overall structural integrity of the head domain [9]. Thus this mutation is thought to affect severely the function of the protein. The second allele carries a mutation resulting in another amino acid substitution, at residue 1602, in which a glutamine replaces an arginine. Some patients homozygous for this mutation have the severe USH1B [10]. Thus we have a problem in interpreting the basis of the deafness in the compound heterozygote siblings with atypical Usher syndrome. If the transcript from the first allele is non-functional and homozygosity for the second mutant allele produces the more severe USH1B, why do these patients not also acquire the more severe form of Usher syndrome? Liu et al. [8] suggest that the genetic background of these families may be a contributing factor to the severity of the disease. The importance of this extends beyond the clinical manifestation of the disease as it suggests that other proteins may affect and/or be affected by myosin VIIa, or that the function of myosin VIIa can be compensated for by other proteins. Identifying these other proteins will be important in understanding the function of myosin VIIa *in vivo* and, potentially, in treating diseases caused by abnormal myosin-VIIa function.

Mouse models

Myosin VIIa was identified as the gene involved in deafness at the shaker1 (*sh1*) locus in mice (*Myo7a*) [11]. The *sh1* mutation arose spontaneously and was first described by Lord and Gates in 1929 [12]. Since then, a further nine *sh1* alleles have been found, and a total of seven mutations have been identified

---

Table 1 (contd.)

Usher syndrome. The mutations are given in the standard nomenclature for human gene mutations [28]; this information can be found on the worldwide web (http://interscience. wiley.com/jpages/1059–7794/nomenclature.html). The number prior to a deletion (del) is the number of the nucleotide deleted and a letter after it shows which type of nucleotide has been deleted (A, T, G or C). The type of deletion is given in parentheses after the nucleotide affected; FS, frame shift (if a number follows, there is a stop codon that number of amino acids, or aa, downstream); IF, in-frame deletion. The number before an insertion (ins) is the number of the affected amino acid and after 'ins' is information describing the exact nature of the insertion. For nonsense and missense mutations, the first three letters are the amino acid affected, followed by its position in the protein, then the amino acid that results from the mutation (or Stop). In the case of intronic mutations, the number directly after IVS (intervening sequence) is the intron affected and the characters following IVS describe where the mutation is in the inton with respect to the splice-acceptor site. Regions of the protein affected by mutations are indicated by: IQ, IQ domain; CC, coiled-coil domain; Talin, talin homology domain. Data are from [1,3,6–8,10,16].

(Table 2). The *sh1* mouse lines are characterized by hearing loss and vestibular defects. To date, none of the *Myo7a* alleles examined has been shown to lead to any signs of retinal degeneration [5]. The *sh1* mice are a useful model for inner-ear abnormalities in humans that are caused by myosin-VIIa dysfunction and are also useful for determining the role of myosin VIIa *in vivo*.

Experimental analysis of the shaker1 mouse has provided great insight into the biological function of myosin VIIa in the inner ear and into the pathology caused by its disruption. However, before we address the possible biological roles of myosin VIIa in hearing and balance, it is important that we understand which cells in the auditory system express myosin VIIa and its intracellular location. We can use an understanding of the location of myosin VIIa within a cell in conjunction with phenotypic analysis of the *sh1* mice to better understand the biological function of myosin VIIa.

Expression and localization of myosin VIIa

In the adult vertebrate inner ear, myosin VIIa is expressed exclusively in the hair cells [13,14] and, during development, is expressed extremely early, prior to stereocilia formation [15]. Antibodies raised against myosin VIIa localize the protein at the light- and electron-microscopic levels within vertebrate hair cells, diffusely spread throughout the cytoplasm but concentrated in several regions, such as the cuticular plate, the pericuticular necklace and within the stereocilia of the hair bundles (Figure 2) [13,14]. Interestingly, the distribution of myosin VIIa within the stereocilia is not the same in all species [13]. In the bullfrog, myosin VIIa is concentrated in a band towards the base of the stereocilia, whereas in mammals myosin VIIa appears to be distributed throughout the stereocilia. This discrepancy in myosin VIIa localization in the stereocilia corresponds to a difference in the distribution of side links: in bullfrogs, side links are concentrated towards the basal end of the stereocilium, whereas in mammals they are distributed evenly along the length of the stereocilium. This interesting correlation implies a role for myosin VIIa in side-link formation and/or maintenance. The localization of myosin VIIa in the cuticular plate, the structure within the hair cell that anchors the stereocilia, suggests that myosin VIIa may be involved in anchoring the stereocilia. Thus the localization of myosin VIIa within the hair cell suggests that it is involved in maintaining the structural integrity of the hair bundle. This hypothesis can be tested by analysing the hair cells of mice with mutations in myosin VIIa.

Lessons from shaker1 mice

The 10 different *Myo7a* alleles identified are all recessive mutations and the exact mutations are known for seven alleles (Table 2). These different mutations affect to some degree the amount of myosin VIIa expressed [5] and its function (Table 2). The 10 different alleles provide the researcher with tools with which to investigate the physiological function of myosin VIIa (for a more complete discussion of myosin VIIa alleles, refer to Mburu et al. [16]). Numerous studies

Table 2. Analysis of the 10 different shaker1 alleles

Allele	Mutation	Protein domain	Protein level	Cochlear physiology	Hair-cell development
sh1	Missense, Arg502→Pro	Head	0.93	Some responses	Normal
6J	Missense, Arg2241→Pro	Head	0.21	Very few responses	Abnormal
26SB	Missense, Phe1800→Ile	Tail	0.46*	No activity	Abnormal
816SB	Intronic, del aa 646–655	Head	0.063	No activity	Abnormal
4494SB	Intronic, Stop	Head	0.0089	Very few responses	Abnormal
4626SB	Nonsense, Gln720→Stop	Head	0.0072	No activity	Abnormal
3336SB	Nonsense, Cys2182→Stop	Tail	0.13	No activity	Abnormal
7J, 8J and 9J	?		?	?	?

Note that for all alleles studied to date, the animals eventually become profoundly deaf. Alleles 7J, 8J and 9J have only recently been identified and the mutations are unknown (?); these mice are deaf and have vestibular abnormalities. Protein levels were determined from kidney and testes and are expressed as a proportion of the wild-type level (taken from [5]). $Myo7a^{26SB}$ is the only allele that gave different results for the protein levels in the two tissues, the value for the testes is shown (*) and the kidney is 0.18, suggesting that there is a differential tissue response to the mutation in $Myo7a$. Cochlear physiology and hair-cell development are from [16]. Deletion of amino acids is indicated by del aa.

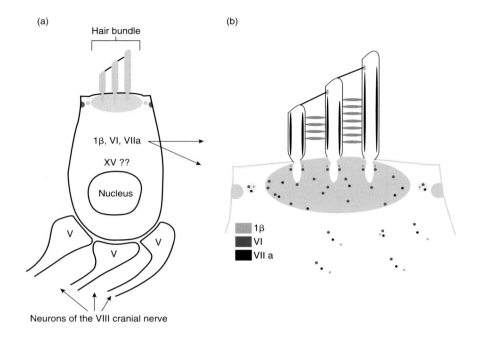

Figure 2. Localization of unconventional myosins in the mammalian hair cell
(a) There are at least five different unconventional myosins expressed in the inner ear (indicated by Roman numerals). The location of myosin XV is not known. Myosin V is located in the postsynaptic terminals of the neurons that synapse with the hair cells. (b) Myosins Iβ, VI and VIIa are all expressed within the hair cell but have unique, though overlapping, expression patterns within the apical portion of the hair cell.

over the last 70 years have analysed the effect of mutations in myosin VIIa on the function of the inner ear with most of the studies concentrating on the original shaker1 allele, $Myo7a^{sh1}$ [12]. The $Myo7a^{sh1}$ mutation is a missense mutation within the part of the gene encoding a poorly conserved region of the myosin head domain [11] and so may have only subtle effects on the function of the head [16]. $Myo7a^{sh1}$ mutants do produce myosin VIIa; in fact, in some tissues it reaches nearly normal levels [5]. Even so, these animals do eventually become deaf and the hair cells in the cochlea degenerate [17].

In a detailed study, electron microscopic and electrophysiological techniques were used to analyse the effect that $Myo7a^{sh1}$ has on the structure and function of the developing hair cells of the cochlea [18]. Ultrastructurally, hair-cell development appears to be approximately normal for the first few weeks after birth, with the exception that only two rows of stereocilia are in the hair bundle instead of the normal three. Not surprisingly, young $Myo7a^{sh1}$ mutants do have some electrophysiological function remaining, although the physiology of the hair cells is not normal. The analysis of the $Myo7a^{sh1}$ mice suggests that myosin VIIa is necessary for maintaining hair-cell viability (they eventually degenerate); in particular, it plays a role in stereocilia development [18].

Self et al. [18] also analysed another allele, $Myo7a^{816SB}$, with more severe effects. These mutants have very low protein levels [5] and any residual protein is thought to be non-functional [16]. The mice have an extensive disruption of hair-cell development in the cochlea, characterized by severely disorganized hair bundles. Furthermore, the $Myo7a^{816SB}$ mutants have no electrophysiological response to sound. Thus the absence of myosin VIIa leads to a severe disruption of hair-bundle development and hair-cell function.

Potential roles for myosin VIIa

By correlating the intracellular localization of myosin VIIa and the phenotypes of the mouse mutants we can begin to elucidate myosin VIIa function *in vivo*. Both its localization near side links and within the cuticular plate suggest that it is involved in maintaining the structural integrity of the hair bundle and that when myosin VIIa is mutated the hair bundle is severely disorganized. Thus myosin VIIa's function in the hair cell appears to be to maintain the proper position of the stereocilia.

Myosin VI

Mutations in the mouse myosin VI gene (*Myo6*) are responsible for deafness and vestibular dysfunction in the Snell's waltzer mouse [19]. To date, no mutations in the human gene for myosin VI (*MYO6*) are known to cause deafness; however, several deafness loci map near the *MYO6* locus [20]. Myosin VI is predicted to be 1266 amino acids in length. Its motor domain contains one ATP-binding domain and one actin-binding domain, and its tail has one IQ motif and a coiled-coil region. Two Snell's waltzer alleles have been identified; both are recessive alleles and both lead to hair-cell degeneration. One mutant line, se^{sv}, is the result of an inversion that has disrupted the function of at least one other gene; in fact, the mutant is named after the other locus that is disrupted—the short-ear locus. The inversion does not disrupt the coding sequence of *Myo6*, but does affect regulatory regions of the gene; se^{sv} mutants have normal-size RNA transcripts but myosin-VI protein levels are only about 15% of those in controls [19]. The second mutation, *sv*, is a deletion that includes 130 bp of coding sequence, resulting in a termination codon truncating the protein just after the head region. There is no detectable myosin VI and this deletion is effectively a null mutation.

Expression and localization

In the ear, myosin VI expression is, in some respects, similar to the expression of myosin VIIa; both are expressed during the earliest stages of hair-cell differentiation [15] and in the adult mammalian hair cell [13]. Furthermore, myosin VI, like VIIa, is concentrated in the cuticular plate and in the pericuticular necklace [13]. However, in contrast with myosin VIIa, myosin VI was not present in the stereocilia of any vertebrate examined [13].

Lessons from Snell's waltzer mice

The phenotype of the *sv* mutant mouse is similar to the *Myo7a^{sh1}* mouse in that there is a severe disorganization of the stereocilia; however, the exact nature of the disorganization is quite different to that of the *Myo7a^{sh1}* mouse. The stereocilia of the *sv* mutants become fused together, resulting in either a single stereocilium or a few large stereocilia per hair cell [20a]. By the time hair cells would normally be morphologically mature, many hair cells in Snell's waltzer mutants are dead or dying and the sensory epithelium of the cochlea eventually degenerates.

Potential roles for myosin VI

The expression pattern of myosin VI in hair cells is similar to that of myosin VIIa and the phenotype of myosin VI-deficient hair cells suggests that myosin VI is necessary for the proper development of the stereocilia. T. Self et al. [20a] suggest that myosin VI directly or indirectly anchors the apical membrane between the stereocilia, preventing the surface tension (resulting from the stereocilia protrusions) from 'zipping up' the stereocilia into one large stereocilium, the exact phenotype of the *sv* mouse. This idea is supported by experiments on *Drosophila* syncytial blastoderms where the injection of antibodies against myosin VI prevented the furrow canals (inward extensions of the membrane) from forming completely [21]. Although the furrow canals are in the opposite direction to the stereocilia, they still must overcome the surface tension created as they are drawn further and further into the cell. Myosin VI function may be to provide the motor that allows some cellular extensions to form normally.

Myosin XV

Myosin XV is the most recent member of the myosin superfamily to be identified in mammals. It was identified as the gene responsible for a non-syndromic recessive deafness (*DFNB3*) [22] and for the shaker2 (*sh2*) mouse phenotype [23]. The myosin XV gene in humans (*MYO15*) has 50 exons and its longest open reading frame found to date is 4757 bp. Structurally, the head domain of myosin XV appears to contain one ATP-binding site and two putative actin-binding sites. The tail region has two IQ motifs, one MyTH4 domain, and one talin-like domain towards the C-terminus. Three different mutations in *MYO15* have been identified that lead to deafness in humans [22]. One is a nonsense mutation in exon 39 that is predicted to result in a truncated protein lacking the talin-like domain. The two other mutations identified by Wang et al. [22] are both single-base transversions causing amino acid substitution within the MyTH4 domain. In fact, the affected amino acids are separated by only two residues. This suggests that the MyTH4 domain may be an important domain for auditory function and analysis of these mutants may shed light on the function of this domain.

Expression and localization

Unfortunately, as *MYO15* has only just been cloned, there are no antibodies available to establish where in the hair cell the protein is. In fact, it is not known whether myosin XV is expressed in the hair cell at all, although it is extremely likely that it is. It will be important to determine the precise localization of myosin XV in order to establish its function in the inner ear.

Lessons from shaker2 mice

The *sh2* mouse is profoundly deaf and has vestibular abnormalities. The stereocilia of *sh2* hair cells are arranged normally but are abnormally short, and the actin cytoskeleton within the cell body is malformed [23]. *sh2* mice have large ectopic actin bundles that stretch from the cuticular plate to the base of the hair cell. This abnormal actin structure suggests that myosin XV is necessary for normal actin/cytoskeleton organization. The mutation in *Myo15* in the *sh2* mouse lies within one of the putative actin-binding domains; it is a single-base transition that produces a cysteine-to-tyrosine (non-conservative) substitution at codon 674. The cysteine at this position is conserved in the vast majority of myosin heads. Thus this residue may be instrumental in actin binding.

Potential roles for myosin XV

As *MYO15* has only recently been cloned and there is little known about myosin XV, it may be premature to speculate on myosin XV's role in the ear. However, based on the ultrastructure of the *sh2* hair cells, it appears that myosin XV functions, in some way, to organize the actin cytoskeleton in hair cells. It will be important to determine if myosin XV's temporal and spatial expression patterns correlate directly with this hypothesis.

Other myosins involved in hearing

There are several other unconventional myosins expressed in the vertebrate inner ear and they may play important roles in inner-ear physiology; however, to date no inner-ear abnormalities in either humans or mice have been identified that are the result of mutations in their genes. Myosin X is expressed within the inner ear, but its localization is unknown [24]. Myosin V is expressed by the neurons of the VIII cranial nerve and is localized in the nerve terminals that synapse on to hair cells. There are several mouse mutants with mutations in myosin V but none of them have been reported to display any inner-ear abnormalities, suggesting that myosin V may not be critical for hearing and balance. At least four different type-1 myosins are expressed in the mammalian inner ear [24,25]. In vertebrates, myosin 1β is expressed in the cuticular plate and stereocilia of hair cells. Myosin 1β in the stereocilia is of particular interest because it is clustered in and around the area where the tip links are anchored (Figure 1b) [13,26,27]. Tip links are extracellular attachments between the tip of a stereocilium to the side of its taller neighbour. The tension of the tip links is thought to be important in the mechanical adaptation of the hair cells and the motor molecule responsible for this is hypothesized to reside in the tip-link anchor of the taller stereocilia. Thus myosin 1β's localization in the tip-link anchoring plaque makes it an ideal candidate to be the motor molecule, or at least one of the motor molecules, responsible for adaptation (Figure 2). To test this hypothesis it will be necessary to disrupt myosin-1β expression within the hair cell. As there is an ever-growing number of unconventional myosins, it is likely that even more myosin isozymes will be found to play important roles in the inner ear and its hair cells.

Perspectives

There is still much to do before we understand the role of myosins in cochlear function. In particular, the difficulty in expressing the unconventional myosin molecules *in vitro* has hampered research into the biochemistry and biophysics of these important motor molecules. The role of myosin 1β may be critical to hair-cell function, located as it is at the heart of the mechano-electrical transduction mechanism, but as yet no mutations have been reported that might allow a critical investigation of the role of this molecule. Although this review has focused upon the role of myosins in deafness, myosin VIIa at least also has a role in the maintenance of retinal function, shown by the retinitis pigmentosa of Usher syndrome. As this is progressive, there seems to be a reasonable prospect of stopping its progression if we only understood the molecular basis of the disease. Progressive hearing loss is extremely common in the human population, and the involvement of mild mutations of the myosin genes described here have yet to be uncovered.

Summary

- *The proper expression and function of several unconventional myosins are necessary for inner-ear function. Mutations in MYO7A and MYO15 cause deafness in humans, and mice. Whereas mutations in Myo6 cause inner-ear abnormalities in mice, as yet no human deafness has been found to the result of mutations in MYO6.*
- *In the mammalian inner ear there are at least nine different unconventional myosin isozymes expressed. Myosin 1β, VI, VIIa and probably XV are all expressed within a single cell in the inner ear, the hair cell.*

- *The myosin isozymes expressed in the hair cell all have unique domains of expression and in some areas, such as the periculticular necklace, several domains overlap. This suggests that these myosins all have unique functions and that all are individually targeted within the hair cell.*
- *The mouse is proving to be a useful model organism for studying both human deafnesses and elucidating the normal functions of unconventional myosins in vivo.*

The authors would like to thank Dr. Amy Kiernan for her helpful discussions and critical reading of the manuscript. The work was supported by the MRC, Defeating Deafness and European Commission contracts BMH4-CT96-1324 and BMH4-CT97-2715.

References

1. Weil, D., Blanchard, S., Kaplan, J., Guilford, P., Gibson, F., Walsh, J., Mburu, P., Varela, A., Levilliers, J., Weston, M.D. et al. (1995) Defective myosin VIIA gene responsible for Usher syndrome type IB. *Nature (London)* **374**, 60–61

2. Kelley, P.M., Weston, M.D., Chen, Z.Y., Orten, D.J., Hasson, T., Overbeck, L.D., Pinnt, J., Talmadge, C.B., Ing, P., Mooseker, M.S., Corey, D., Sumegi, J. & Kimberling, W.J. (1997) The genomic structure of the gene defective in Usher syndrome type Ib (MYO7A). *Genomics* **40**, 73–79

3. Weil, D., Kussel, P., Blanchard, S., Levy, G., Levi-Acobas, F., Drira, M., Ayadi, H. & Petit, C. (1997) The autosomal recessive isolated deafness, DFNB2, and the Usher IB syndrome are allelic defects of the myosin-VIIA gene. *Nat. Genet.* **16**, 191–193

4. Wolfrum, U., Liu, X., Schmitt, A., Udovichenko, I.P. & Williams, D.S. (1998) Myosin VIIa as a common component of cilia and microvilli. *Cell Motility Cytoskel.* **40**, 261–271

5. Hasson, T., Walsh, J., Cable, J., Mooseker, M.S., Brown, S.D. & Steel, K.P. (1997) Effects of shaker-1 mutations on myosin-VIIa protein and mRNA expression. *Cell Motility Cytoskel.* **37**, 127–138

6. Liu, X.Z., Walsh, J., Tamagawa, Y., Kitamura, K., Nishizawa, M., Steel, K.P. & Brown, S.D. (1997) Autosomal dominant non-syndromic deafness caused by a mutation in the myosin VIIA gene. *Nat. Genet.* **17**, 268–269

7. Liu, X.Z., Walsh, J., Mburu, P., Kendrick-Jones, J., Cope, M.J., Steel, K.P. & Brown, S.D. (1997) Mutations in the myosin VIIA gene cause non-syndromic recessive deafness. *Nat. Genet.* **16**, 188–190

8. Liu, X.Z., Hope, C., Walsh, J., Newton, V., Ke, X.M., Liang, C.Y., Xu, L.R., Zhou, J.M., Trump, D., Steel, K.P. et al. (1998) Mutations in the myosin VIIA gene cause a wide phenotypic spectrum, including atypical Usher syndrome [letter]. *Am. J. Hum. Genet.* **63**, 909–912

9. Cope, M.J.T., Whisstock, J., Rayment, I. & Kendrick-Jones, J. (1996) Conservation within the myosin motor domain: implications for structure and function. *Structure* **4**, 969–987

10. Weston, M.D., Carney, C.A., Rivedal, S.A. & Kimberling, W.J. (1998) Spectrum of myosin VIIa mutations causing Usher syndrome type Ib. *Assoc. Res. Otolaryngol. Abstr.* **21**, 41

11. Gibson, F., Walsh, J., Mburu, P., Varela, A., Brown, K.A., Antonio, M., Beisel, K.W., Steel, K.P. & Brown, S.D. (1995) A type VII myosin encoded by the mouse deafness gene shaker-1. *Nature (London)* **374**, 62–64

12. Lord, E.M. & Gates, W.H. (1929) Shaker, a new mutation of the house mouse (*Mus musculus*). *Am. Nat.* **63**, 435–442

13. Hasson, T., Gillespie, P.G., Garcia, J.A., MacDonald, R.B., Zhao, Y., Yee, A.G., Mooseker, M.S. & Corey, D.P. (1997) Unconventional myosins in inner-ear sensory epithelia. *J. Cell Biol.* **137**, 1287–1307

14. Hasson, T., Heintzelman, M.B., Santos-Sacchi, J., Corey, D.P. & Mooseker, M.S. (1995) Expression in cochlea and retina of myosin VIIa, the gene product defective in Usher syndrome type 1B. *Proc. Natl. Acad. Sci. U.S.A.* **92**, 9815–9819

15. Xiang, M., Gao, W.Q., Hasson, T. & Shin, J.J. (1998) Requirement for Brn-3c in maturation and survival, but not in fate determination of inner ear hair cells. *Development* **125**, 3935–3946

16. Mburu, P., Liu, X.Z., Walsh, J., Saw, Jr., D., Cope, M.J., Gibson, F., Kendrick-Jones, J., Steel, K.P. & Brown, S.D. (1997) Mutation analysis of the mouse myosin VIIA deafness gene. *Genes Funct.* **1**, 191–203

17. Deol, M.S. (1956) The anatomy and development of the mutants pirouette, shaker-1 and waltzer in the mouse. *Proc. R. Soc. Lond. B* **145**, 206–213

18. Self, T., Mahony, M., Fleming, J., Walsh, J., Brown, S.D. & Steel, K.P. (1998) Shaker-1 mutations reveal roles for myosin VIIA in both development and function of cochlear hair cells. *Development* **125**, 557–566

19. Avraham, K.B., Hasson, T., Steel, K.P., Kingsley, D.M., Russell, L.B., Mooseker, M.S., Copeland, N.G. & Jenkins, N.A. (1995) The mouse Snell's waltzer deafness gene encodes an unconventional myosin required for structural integrity of inner ear hair cells. *Nat. Genet.* **11**, 369–375

20. Avraham, K.B., Hasson, T., Sobe, T., Balsara, B., Testa, J.R., Skvorak, A.B., Morton, C.C., Copeland, N.G. & Jenkins, N.A. (1997) Characterization of unconventional MYO6, the human homologue of the gene responsible for deafness in Snell's waltzer mice. *Hum. Mol. Genet.* **6**, 1225–1231

20a. Self, T., Sobe, T., Copeland, N.G., Jenkins, N.A., Avraham, K.B. & Steel, K.P. (1999) Role of myosin VI in the differentiation of cochlear hair cells. *Dev. Biol.* **214**, 331–341

21. Mermall, V. & Miller, K.G. (1995) The 95F unconventional myosin is required for proper organization of the *Drosophila* syncytial blastoderm. *J. Cell Biol.* **129**, 1575–1588

22. Wang, A., Liang, Y., Fridell, R.A., Probst, F.J., Wilcox, E.R., Touchman, J.W., Morton, C.C., Morell, R.J., Noben-Trauth, K., Camper, S.A. & Friedman, T.B. (1998) Association of unconventional myosin MYO15 mutations with human nonsyndromic deafness DFNB3. *Science* **280**, 1447–1451

23. Probst, F.J., Fridell, R.A., Raphael, Y., Saunders, T.L., Wang, A., Liang, Y., Morell, R.J., Touchman, J.W., Lyons, R.H., Noben-Trauth, K., Friedman, T.B. & Camper, S.A. (1998) Correction of deafness in shaker-2 mice by an unconventional myosin in a BAC transgene. *Science* **280**, 1444–1447

24. Solc, C.K., Derfler, B.H., Duyk, G.M. & Corey, D.P. (1994) Molecular cloning of myosins from bullfrog saccular macula: a canidate for the hair-cell adaptation motor. *Auditory Neurosci.* **1**, 63–75

25. Crozet, F., el Amraoui, A., Blanchard, S., Lenoir, M., Ripoll, C., Vago, P., Hamel, C., Fizames, C., Levi-Acobas, F., Depetris, D., Mattei, M.G. et al. (1997) Cloning of the genes encoding two murine and human cochlear unconventional type I myosins. *Genomics* **40**, 332–341

26. Steyger, P.S., Gillespie, P.G. & Baird, R.A. (1998) Myosin I beta is located at tip link anchors in vestibular hair bundles. *J. Neurosci.* **18**, 4603–4615

27. Garcia, J.A., Yee, A.G., Gillespie, P.G. & Corey, D.P. (1998) Localization of myosin I beta near both ends of tip links in frog saccular hair cells. *J. Neurosci.* **18**, 8637–8647

28. Antonarakis, S.E. & Nomenclature Working Group (1998) Recommendations for a nomenclature system for human gene mutations. *Hum. Mut.* **11**, 1–3